£ 16

HAWORTHIA
AND ASTROLOBA
A Collector's Guide

HAWORTHIA
AND ASTROLOBA
A Collector's Guide

John Pilbeam

Photography by Bill Weightman

B. T. BATSFORD LTD · LONDON

ISBN 0 7134 0534 1

Typeset by Keyspools Ltd, Golborne, Lancs.
and printed in Great Britain by
Butler & Tanner Ltd. Frome, Somerset
for the publishers
B. T. Batsford Ltd.
4 Fitzhardinge Street
London W1H 0AH

CONTENTS

Acknowledgments   7
List of Colour Plates   7
Introduction   9

Cultivation   11
    Growing medium—Watering—
    Shading—Temperature—Vegetative
    propagation—Growing from seed—
    Variegates—Pests and diseases

Classification: Revision of Sections   14

Flowers, Fruit and Seed   16

Distribution of Haworthia Species   21

Checklist of Species   31

Commentary on Species   33

Astroloba   145
    Cultivation—Classification—
    Distribution—Checklist of Species—
    Commentary on Species

Glossary of Terms   155
Societies   159
Bibliography   161
Index   163

ACKNOWLEDGMENTS

I should like to thank the following in particular for their help during the preparation of this book:

Bill Weightman, of Orpington, Kent, whose excellent and painstaking photography shows the beauty of these plants so well; Mel Roberts, of New Ferry, Wirral, for the paintings illustrating some of the species, which add another dimension to their appreciation; Bruce Bayer, of the Karoo Botanic Garden, South Africa, for a wealth of information in correspondence over the last ten years or so, for plants sent to facilitate study, and for some of the photographs in this book; Peter Brandham, of the Jodrell Laboratory, Royal Botanic Gardens, Kew, for allowing me access to study and photograph plants in his care, largely resulting from his collecting trips; Derek Tribble, for help with the distribution maps, and provision of some of the photographs; Susan Holmes of the Royal Botanic Gardens, Kew, for supplying the necessary Latin diagnoses in the Classification chapter (verified by her colleague, Mr H. K. Airy-Shaw); Jack Brown, of Uxbridge, Middlesex, for early encouragement and generous gifts of plants from his collection; Geoff Francis for producing the main map; Jane Swanbrow, for typing nobly from the original impossible manuscript; all the other Haworthia enthusiasts who have encouraged me to continue with this work, in spite of setbacks.

LIST OF COLOUR PLATES

PLATE 1
H. cymbiformis var. *cymbiformis* (painting by Mel Roberts)
H. limifolia var. *striata* (painting by Mel Roberts)
H. venosa subsp. *tessellata* (painting by Mel Roberts)

PLATE 2
H. archeri var. *archeri*
H. attenuata fa. *caespitosa*
H. bolusii var. *blackbeardiana*
H. bruynsii
H. coarctata (*greenii*)
H. coarctata subsp. *coarctata* var. *tenuis*
H. coarctata subsp. *adelaidensis* fa. *adelaidensis*
H. comptoniana fa. *major*

PLATE 3
H. cooperi var. *cooperi* fa. *pilifera*
H. cymbiformis var. *incurvula*
H. cymbiformis (variegate)
H. decipiens
H. divergens
H. emelyae var. *emelyae* from Van Wyksdorp
H. floribunda from Riversdale
H. habdomadis var. *habdomadis*

PLATE 4
H. herbacea
H. koelmaniorum
H. limifolia var. *gigantea*
H. lockwoodii from Laingsburg
H. maculata from Worcester
H. magnifica var. *magnifica* from Heidelberg
H. magnifica var. *major* from Riversdale
H. magnifica var. *maraisii* from Robertson

PLATE 5
H. magnifica var. *meiringii* from Bonnievale
H. marginata
H. minima
H. mirabilis subsp. *mirabilis* fa. *beukmannii*
H. mirabilis subsp. *mirabilis* fa. *napierensis*
H. mirabilis subsp. *mirabilis* fa. *rubrodentata*
H. mirabilis subsp. *badia*
H. mirabilis subsp. *mundula* from Muiskraal

PLATE 6
H. mutica from Bredasdorp
H. nigra fa. *angustata* from Thomas River
H. nortieri var. *nortieri*
H. nortieri var. *globosiflora* from Botterkloof

H. parksiana
H. poellnitziana
H. pumila
H. pygmaea fa. *major* from Great Brak
PLATE 7
H. reinwardtii var. *brevicula*
H. reinwardtii var. *reinwardtii* fa. *chalumnensis*
H. reinwardtii var. *reinwardtii* fa. *kaffirdriftensis*
H. reinwardtii var. *reinwardtii* fa. *zebrina*
H. reticulata var. *hurlingii* near Bonnievale
H. retusa var. *retusa* fa. *retusa* from Albertinia

H. retusa var. *acuminata* from Gouritz
H. retusa var. *dekenahii*
PLATE 8
H. rycroftiana
H. semiviva from Frazerburg
H. serrata
H. subattenuata (variegate)
H. truncata fa. *truncata*
H. ubomboensis
H. venosa subsp. *venosa* by the Breede River
Astroloba foliolosa

All the drawings are by the author.

INTRODUCTION

The literature on *Haworthia*, this genus of succulent plants from South Africa, has been hitherto scattered, mainly in the form of notes or the naming of species in often obscure publications in various languages. It quickly becomes apparent in studying Haworthias that there is a great deal of confusion about the naming of the plants in cultivation, not in any way helped by the frequent misidentification in popular books on succulents which tend to be the stand-by for the collector. Reference to original descriptions of species is essential for any detailed study and understanding, and reveals many such anomalies, as well as creating a few.

Some of the botanists who have concerned themselves with this complex genus must take part of the blame, as they have often based their identifications on single plants, sometimes not typical and often unrelated to field populations. Illustrations accompanying descriptions of new species and a proliferation of varieties have often been inadequate for identification or absent altogether, which in view of the complexity of the genus makes real determination of what was described impossible with certainty.

What few general listings of species there have been are similarly devoid of sufficient illustration, being merely collections of published names without real information on the genus as a whole. The present situation is that Jacobsen's listing in his *Handbook of Succulent Plants* and his *Succulent Lexicon* are still the most often-used guides to the species in this genus.

It is with joy that the researcher, delving into the literary references to the genus, stumbles into the first daylight in the form of Salm-Dyck's monograph on Aloes and Mesembryanthemaceae compiled over a century ago, with beautiful, dextrously executed drawings. The name implies that this monograph covers only the genus of *Aloe* in the family Liliaceae; but in addition to covering that genus, because Salm-Dyck did not apparently accept some of the hiving off from *Aloe* that had been taking place, it also covers *Gasteria*, *Haworthia* and *Astroloba*, the last referred to as *Apicra*, a later invalidated name.

G. G. Smith too, working in the field in the 1930s and 1940s, recorded his work well, with photographs accompanying all his descriptions in the *Journal of South African Botany*. And M. B. Bayer's papers over the last few years have been accompanied by clear photographs leaving no doubt as to identification. More useful generally than any other large-scale work on the genus has been John R. Brown's 'Notes on Haworthias' in the American Society's journal, published intermittently over the last 40 years, with excellent photographs and references in detail to past literature.

Apart from Berger's section in Engler's *Pflanzenreich* in 1908 and Jacobsen's efforts at cataloguing the species in his *Handbook of Succulent Plants* and his *Succulent Lexicon*, the only collected comprehensive reference to this genus under one cover is Bayer's *Haworthia Handbook*, published in 1976, which examines and completely revises the taxonomy of *Haworthia*. Bayer's checklist in his *Handbook* is indeed the basis on which this book is written, and the writer welcomed the revision, which was long overdue. Based as it is on extensive field work and intensive study of plants both in habitat and under cultivation at the Karoo Botanic Garden in Worcester, Cape Province, South Africa, it is an authoritative and important work in the understanding of the genus.

The New Haworthia Handbook, by Bayer, a revision of his 1976 work, has recently been published, and is a worthy follow-up to it, with expanded notes to the species and colour photographs illustrating them.

Having grown many collected plants sent by Bruce Bayer over the last several years I am in full agreement (well, very nearly) with his revisions. However I am loath to discard altogether accepted names for some of the forms of very variable species, and have retained some at form level for distinct collectors' plants, with his concurrence. This variation is shown too by several illustrations of some species, variation not always warranting even form recognition, else we would be back where we started before Bayer's work.

In addition an attempt has been made in the present work to reshape the sections of *Haworthia* to take account of the revised concept of the relationships of species and to assist in identification.

Many of the plants received from habitat show the wide variation of species as evidenced so well in Bayer's articles, which have appeared in the National Cactus & Succulent Society's journal, published in England (Vol. 27:10 (1972) and 28:80 (1973)). These papers have helped to engender an understanding of the variability of species, which has led in the past to the oversplitting of species to an unacceptable and sometimes ridiculous degree.

Various students of the genus have had a monograph in mind as they stepped deeper and deeper into the morass of names, but for one reason or another none appeared before 1976. For this reason and because Bayer's work is very much a botanist's work, leaving many of the needs of the collector unprovided for, you are now reading this attempt at bringing some order and sense comprehensible to collectors of this beautiful genus of succulent plants.

The task of producing such a work has necessarily involved hours of delving into obscure references, and appraisal of a comprehensive collection of plants built up over a period of about 20 years. This would have been impossible without co-operation from someone in the field, prepared to study in habitat and correspond freely and enthusiastically. I have received such help unstintingly from Bruce Bayer, with whom I have corresponded for the last ten years. Without his help and encouragement I should have been unable to complete this book.

The plants themselves collected together and grown over this time reveal their affinities in various ways, not least by comparison of the flowers and seeds, which sometimes confirm superficial resemblances and sometimes surprisingly deny them. Growing batches of habitat-collected seed reveals the extent of variation within a species, which sometimes (as in *H. emelyae*) extends to superficially quite different-looking plants. From many plants received from habitat together with enlightening letters from Bruce Bayer, it is apparent

that species of *Haworthia* show often a wide degree of variation within the species, and sometimes refuse to keep themselves to themselves, i.e. they hybridize. This has been ably set forth in the National Cactus & Succulent Society's journal, in the previously mentioned articles. Bayer's work involves a drastic reduction in the number of species, welcomed by most collectors. To make clear what has been submerged and where, I have included reference to former species (and lesser taxa where appropriate) in the main text.

New names published in this book:
Haworthia comptoniana fa. *major* fa. nov.
Haworthia limifolia var. *striata* var. nov.
Haworthia pygmaea fa. *crystallina* fa. nov.
Haworthia pygmaea fa. *major* fa. nov.

CULTIVATION

The situation of these plants in habitat varies enormously: from completely exposed species in sparse vegetation to others which grow almost completely out of view of the sun. The majority grow with some regard for avoiding the worst excesses of the sun in South Africa, by secreting themselves among grasses and bushes, so that they receive partial shade, although it should be remembered that with sparse, sun-dried foliage on the protecting shrubs and intense sunlight this still means a high sun dosage compared with that received in more temperate zones. A few on the other hand grow in conditions of complete shade from the sun on south-facing slopes (sun-less in the southern hemisphere of course), for example *H. translucens* subsp. *tenera*, which I have seen photographed in such a situation growing among mosses! And others sit with the main part of the plant body buried for protection with just translucent leaf-tips protruding into the light.

Whatever the situation this genus demonstrates the gamut of ways of conserving moisture: some plants, notably in the Sections Fenestratae and Retusae have contractile roots which pull the plants down into the soil, allowing them to emerge further into the light at times of water-plenty; some close their leaves into a tight ball, like a closing water-lily, to protect the inner part of the rosette from the sun, often with thickened areas of tubercles for further protection; others form columns of close-packed leaves and grow into dense clusters of stems or rosettes the better to keep cool the lower plant body; many have attenuated leaves which dry up progressively from the tips as the water content of the plant diminishes, the dried parts themselves closing over the crown to form a protective layer for the heart of the plant; one or two have tuberous and many very thick roots, which act as water-storing organs. Combinations of these devices are often found in the one species.

Growing medium

The nature of these plants as just expounded must indicate cultivation in a porous, open compost with a good proportion of grit or coarse sand. Choice of growing medium is an individual matter, and if you have already found a mixture which seems to suit your plants, then stick to it. For many years I have grown Haworthias in John Innes formula potting composts with very gritty sand from which the dust has been removed by sieving, in the proportions of two parts compost to one part grit, varying the proportions according to the requirements of the individuals; the more difficult or slow-growing species get more grit content. For the last few years, however, I have increasingly used soil-less composts based on peat, similarly opened by adding gritty sand, and at present I am using a mixture of the two, i.e. two parts John Innes No. 2 potting compost, two parts soil-less compost and three parts gritty sand, with a layer of grit on the top, to combat algae and mosses. After a few months there is little nutrient left in the compost and I therefore feed with a high-potash liquid fertilizer about once a month through the growing season.

Watering

Unless the plants are subjected to low temperatures, below about 6°C (40°F), watering can continue all the year round, but easing off during the winter months. However, if a plant is clearly resting watering should be kept down until active growth commences, and with some species there seems to be more resting than growing. These are usually the difficult-to-grow species from arid areas, which need far less water at any time.

Shading

As indicated at the beginning of this chapter Haworthias in the wild occur in a number of different situations as regards the amount of sunlight they receive. But what must not be lost sight of is that we are talking about a country where the intensity of the light and the power of the sun is considerably more than ever occurs in England, for example. They will not thrive therefore if they are grown in positions where they receive little sunshine at all, although they will tolerate such murky positions. In a glass-to-ground glasshouse they will mostly grow quite well under the staging provided there is at least exposure to a few hours' sunshine every day. Some species prefer more exposure, like the species in the Section Retusae, which are adapted to

Spartan conditions well exposed to the elements in the wild. But some of the softer-leaved species, for instance in the Section Arachnoideae, will not take too much sun at all. Change them around and find the right position for each species: if they start looking brown and dried up (although they do colour well in full light, remember), give them a less sunny position; or if they start reaching for the light, give them gradually more exposure. Those growers in more sunny climes will of course take measures more appropriate to preventing their shrivelling in excessive sun, by shading them with lathes or netting.

Temperature

Haworthias will tolerate quite low temperatures, down to nearly freezing point, without damage, but they prefer warmer conditions without doubt; for safety aim at a minimum of 5°C (40°F). The lower the temperature allowed the drier they should be kept. They will not tolerate low temperatures and damp feet as well.

Vegetative propagation

Many Haworthias are easily propagated by removal of the sometimes numerous offsets which appear around the base of the plants. These should be carefully removed with a sharp knife as close as possible to the main rosette in order that any roots already forming on the new shoot's stem may be taken advantage of. As the growing point is often deceptively deep down, this will also prevent taking just the top of the shoot and finishing up with a handful of leaves. The cut surface should be dusted with hormone rooting powder, as much for its fungicidal properties as anything, and left to dry thoroughly before placing after a few days in dry compost. Unless roots are already present, withhold water until new roots can be seen at the base of the cutting. Keep out of strong light, to prevent too much desiccation, until the roots have formed.

Individual leaves of Haworthias may also be rooted to produce one or more small plants. As whole a leaf as possible should be detached, and the damaged surface dusted with hormone powder. After allowing to dry for a day or two place in dry compost, and treat as for cuttings; several shoots are usually sent up by leaves so rooted. Thin-substance leaves are difficult to root, however, and some of those plants with hard, leathery leaves will take a long time to do so.

Offsets are occasionally formed on the flower-stem, especially those species which have branching flower-stems. If kept well watered the stem beneath the offset will remain turgid; keep this adventitious plantlet growing until it reaches the size that you would normally take off for a cutting, when it may be removed and treated as such.

Growing from seed

Haworthias grow readily, and fairly quickly, from seed. If more than one plant of a species is available, and they are not offsets from the same original plant, pollination to produce true offspring can be effected. This is best done away from other plants and away from wandering natural pollinators, like bees, and can be achieved by inserting a thin paintbrush into the flowers of one plant and transferring the pollen to the deep-seated stigma of the other, and then reversing the process. The maturing seed-capsules should be watched carefully, as, when ripe, they will split and shed the seed. This can be prevented by putting a small piece of sticky tape round the capsule. Once the capsules have dried and split the papery seed may be removed and sown with advantage immediately, as it is not long-lived.

A close atmosphere should be maintained until germination has occurred and for a month or two after this, to give the seedlings time to get their roots down into the growing medium and sustain the single cotyledon leaf. Once two or three leaves have developed the seedlings may be given more air and filtered sunlight. On no account should the compost be allowed to dry out, however, in these first months, or for the first year, unless they are to be kept at low temperatures.

A temperature of 21°C (70°F) should be aimed at for germination. After this 10–13°C (50–55°F) will suffice. Regular doses with fungicide and insecticide should be given, especially to combat damping-off fungus and that insidious pest the Sciara fly, whose maggots will dispose of a whole sowing given the chance.

Keep records of your propagations, especially if you have plants of known origin in the wild, and pass on the information to collectors who have your propagations; let us keep the records straight.

Variegation

Apart from the numerous hybrid Haworthias, which cling on in collections, and have their supporters to ensure their survival for at least a few decades more, there are a few other oddities, which have their devotees too.

The most commonly seen are the variegated forms of a few Haworthia species. Variegation is caused by a lack of chlorophyll pigment in part of the leaf, which in various degrees gives stripes of colouring in white, yellow or orange according to the depth of the natural coloration of the species. If the variegation is complete, and wholly white, yellow or orange leaves are produced, the plant will surely die, for without some green colouring it cannot use the light it receives to achieve growth, by means of the process known as photosynthesis carried out by the chlorophyll in the plant. So do not remove such offsets with no green colouring if they are formed on a variegated plant, for without their attachment to the partly green part of the plant they cannot survive.

I know of only five species found with variegation. They are:

H. *cooperi* fa. *pilifera*;
H. *cymbiformis* var. *cymbiformis*;
H. *attenuata* fa. *caespitosa*;
H. cv. *subattenuata*;
H. *tortuosa*.

In the first two mentioned, the variegation is white or cream-coloured, in the latter three, yellow or orange, varying from plant to plant. Occasionally the variegation is obscure and underlying the surface, in which case a pallid, somewhat pink and green striped effect is produced.

Other oddities

There are a few Haworthias which never seem to flower in cultivation, and they are perhaps indicating hybrid origin from other than *Haworthia*. There is a form of *H. reinwardtii* which produces a flower stem which constantly branches and rebranches, never getting around to producing the flowers. This is the nearest thing to a monstrose form in the genus that I know; there are no cristates or monstrose forms to my knowledge.

Pests and diseases

Haworthias suffer very little from pests or diseases, although they are of course subject to attacks by the common scourges of succulent plants, the two forms of mealy-bug. If cottonwool-like patches are seen among the leaves or in the growing point examination will usually reveal the tiny mealy-bugs, like powdered tiny woodlice or pill-bugs. On the roots the evidence of their presence is the same, cottonwool-like patches, with even tinier white, sausage-shaped bugs busily sucking the sap from the roots. Both forms of this pest suck the sap of the plant and the best preventative is the regular use of systemic insecticides, which are taken up by the plants and then poison the pests.

Sciara or mushroom fly is the only other pest which I have suffered from with Haworthias; they can be drastically reduced by not using soil-less composts, which the fly seems to favour. Otherwise regular drenches with contact or systemic insecticides and overhead spraying to knock down the flies seem the best measures.

Some species suffer from black marking, which seems to demolish the tissue, without usually proving fatal, so that parts of leaves are damaged and become unsightly. There seems to be nothing other than physical causes for this disorder, and perhaps improved growing conditions are the best solution. I have found, perhaps coincidentally, that regular spraying with a systemic fungicide seems to combat this disorder, more commonly encountered in the genus *Gasteria*. It remains mysterious.

CLASSIFICATION

From the time Duval set up the genus *Haworthia* in 1809, it has been divided into sections to help identification and acknowledge differences and similarities between the species. Until comparatively recently the division has been based on leaf and rosette characters.

Berger, in 1908, proposed 18 sections, which have continued with little real change to be the basis of division until the present time. But he foresaw a division by floral characters, first put into effect by Uitewaal in 1947 (*Des. Pl. Life*, 19:133 (1947)) followed by Bayer more fully in 1971 (*Cact. Amer.*, 43:157 (1971)).

Inevitably other genera closely related to *Haworthia* have from time to time been associated with the genus even more closely, but they have in general not been accepted as partners under this generic name. The contenders have been *Astroloba* Uitew. (syn. *Apicra* Haw.), *Chortolirion* Berg. and *Poellnitzia* Uitew. The first is held to be separate principally by the different character of the flower (see p. 147); the free segments of the corolla are very short and recurve only slightly in a regular manner, unlike the petals of *Haworthia*, which reflex to a much greater degree, with three, usually longer petals curving downwards together and three, usually shorter petals curving upwards less strongly, but separated into two groups of three, as shown on p. 16.

Chortolirion flowers are similar to *Haworthia*, but the plants are bulbous in character; Rowley has placed them in the Section Fusiformes, but Bayer has rejected this. *Poellnitzia*, a monotypic genus, bears a superficial resemblance to *Astroloba* in the leaves and habit of growth, but the flower (red) is quite different from *Astroloba* or *Haworthia*, and Rowley has recently combined this monotypic species with *Aloe*. *Astroloba* species and *Poellnitzia rubriflora* are shown on pp. 148–154 to show the similarity of the stems to some *Haworthia* species. The *Haworthia* species they most closely resemble are either much more slender-leaved or distinctly trifarious (with their leaves in three vertical columns) while these two genera have generally more substantial leaves in five rows.

Following Bayer's division of the genus into three subgenera according to floral characters (see pp. 16–18) an attempt is made below to rationalize sections

beneath these subgenera. This takes into account the greater understanding of the variability of species elucidated by Bayer in previously mentioned papers.

It must be emphasised that any such man-made division commands no respect from the plants themselves, which often have a tentative foot in one or more sections or subsections. Inasmuch as such division helps determine the identity of *Haworthia* species it serves some purpose, but as Bayer emphasizes in his Handbook 'any plant considered out of its natural habitat and without reference to the population from which it comes, particularly in *Haworthia*, has to be regarded with caution.'

Revision of Sections

SUBGENUS HAWORTHIA: perianth at base triangular or rounded-triangular; type *H. arachnoidea*;

SECTION ARACHNOIDEAE Haw.: not recurved at the leaf-tip, more or less transparent, often with teeth on the margins;

SUBSECTION Arachnoideae: base-colour mid-green, not greatly translucent, often opaque at the leaf-tips with longitudinal green lines beneath the surface, solitary or slowly offsetting, often with teeth on the margins—species: *H. arachnoidea*, *H. aranea*, *H. aristata*, *H. decipiens*, *H. xiphiophylla*;

SUBSECTION Limpidae Berg.: base-colour bluish-green to pale green, upper part of leaf translucent with longitudinal green lines beneath the surface, solitary or slowly offsetting, often with teeth on the margins—species: *H. altilinea*, *H. bolusii*, *H. cooperi*, *H. habdomadis*, *H. lockwoodii*, *H. semiviva*, *H. translucens*;

SUBSECTION Proliferae subsect. nov.: *colore medice usque atroviridi; rosulae parvae (usque 4cm diametro), valde aggregatae; folia apice non vel vix translucentia, margine dentata vel edentata; typus: H. marumiana;* base-colour mid to dark green; small rosettes (to 4cm in diameter), strongly clustering; leaves not or a little translucent at the leaf-ends, with or without teeth on the margins; type: *H. marumiana*;—species: *H. batesiana*, *H. marumiana*;

SUBSECTION Cymbifoliae subsect. nov.: *folia plerumque apaca, viridia usque pallide viridia, superne abrupte rotundata, incrassata, pagina superiore basi concava, pagina inferiore convexa, apice plus minusve translucente saèpe translucido-guttato; typus: H. cymbiformis*; leaves usually opaque, green to pale green, abruptly rounded and thickened in the upper part, upper surface concave at base, lower surface convex, more or less translucent at apex of leaf, often in flecks; type: *H. cymbiformis*;—species: *H. cymbiformis*.

SECTION FENESTRATAE von Poelln.: leaves truncate, the upper surface windowed—species: *H. maughanii, H. truncata;*

SECTION LORATAE(Salm-Dyck)Berg.: leaves narrow, not fleshy, edges often armed with minute teeth;

SUBSECTION Loratae: roots not tuberous—species: *H. angustifolia, H. chloracantha, H. divergens, H. floribunda, H. parksiana, H. pulchella, H. variegata, H. wittebergensis, H. zantnerana;*

SUBSECTION Fusiformes Barker: roots thick, tuberous—species: *H. blackburniae, H. graminifolia;*

SECTION RETUSAE Haw.: upper leaf inflated to form more or less distinct end-area, tip somewhat transparent with green lines;

SUBSECTION Retusae: leaf-tips recurved strongly, and/or end-area well defined—species: *H. comptoniana, H. emelyae, H. heidelbergensis, H. magnifica, H. mirabilis, H. mutica, H. pygmaea, H. retusa, H. springbokvlakensis;*

SUBSECTION Turgidae subsect. nov.: *folia inflata, vix recurva, plerumque ad apicem extremum incurva; typus: H. turgida*; leaves inflated, little recurving, usually incurving at the extreme tip; type: *H. turgida*;—species: *H. archeri, H. herbacea, H. maculata, H. nortieri, H. pubescens, H. reticulata, H. rycroftiana, H. serrata, H. turgida.*

SUBGENUS HEXANGULARES Uitew.: perianth at base hexangular or rounded-hexangular, gradually narrowing to junction with pedicel; type: *H. coarctata*;

SECTION COARCTATAE Berg.: incurving, compressed leaves and columnar stems; or recurving, attenuated leaves, more or less acaulescent;

SUBSECTION Coarctatae: incurving leaves, columnar stems—species: *H. armstrongii, H. coarctata, H. fasciata, H. glauca, H. reinwardtii*;

SUBSECTION Attenuatae subsect.nov.: *folia recurva attenuata; rosulae plerumque acaulescentes, rosulae maturae interdum brevicaules; typus: H. attenuata*: leaves recurving attenuate, rosettes usually acaulescent, mature rosettes sometimes short-stemmed; type: *H. attenuata*:—species: *H. attenuata, H. glabrata, H. longiana, H. radula;*

SECTION TRIFARIAE Haw.: leaves in three prominent, or less prominent, somewhat twisted series, usually strongly trifarious;

SUBSECTION Caulescentes Haw.: columnar stems—species: *H. nigra, H. viscosa*;

SUBSECTION Acaules Haw.: stemless, weakly trifarious—species: *H. scabra, H. smitii, H. sordida, H. starkiana*;

SECTION VENOSAE Berg.: leaves triangular or long-triangular, spreading and recurved—species: *H. koelmaniorum, H. limifolia, H. ubomboensis, H. venosa, H. woolleyi.*

SUBGENUS ROBUSTIPEDUNCULARES (Uitew.) Bayer: perianth at base hexangular or rounded-hexangular, abruptly joined to pedicel; type: *H. pumila* (syn. *H. margaritifera*)

SECTION MARGARITIFERAE Haw.: having large, raised, cartilaginous tubercles or edges on the leaves—species: *H. kingiana, H. marginata, H. minima, H. pumila, H. poellnitziana.*

FLOWERS, FRUIT AND SEED

As explained in the classification chapter, the genus *Haworthia* is divided into three subgenera on the basis of the shape of the flowers.

The subgenus Haworthia (syn. Triangulares) has flowers which in lateral section are triangular, and this shape can be observed without sectioning. They also tend to be wider-petalled and shorter-tubed than in the subgenus *Hexangulares*; the petals are coloured clear white to pinkish, with veining varying from commonly green to grey or brownish-pink, and the petals are widely spreading and regularly spaced within the two planes of the upper and lower three petals. The seed capsule is shortly-elongated, the seed grey to brown and small, roughly as long as it is wide. In conjunction with these floral characters the consistency of the leaves is succulent, with a soft epidermis easily broken or pierced, and a sappy interior.

Flowers of subgenus *Haworthia*

The subgenus *Hexangulares* has flowers hexagonal in section, also clearly observable without sectioning, tapering to join the pedicel. The petals tend to be narrow, and longer than in the subgenus *Haworthia*, coloured greyish-white or brownish, often more bunched together and less recurving in the upper petals, the lower petals longer, recurving strongly and lobed at the ends, and with long flower-tubes. The veining is usually grey to dark brown. The seed capsule is elongated, like an airship, and generally longer than in the subgenus *Haworthia*. The seeds are dark brown or black, longer than broad, and a little winged. The leaves are mostly leathery in texture, fibrous inside and not breaking cleanly easily, sappy only when growing strongly.

Flowers of subgenus *Hexangulares*

The subgenus *Robustipedunculares* has flowers rounded-hexangular in section, flattened at the base of the tube and abruptly joining the pedicel. The recurving parts of the petals are not as long as the other subgenera and recurve weakly and almost regularly, with little difference between the upper and lower three petals. The flowers are greenish-white, with green or grey veining. Seed capsules are large and rounded-elongate. Seeds are dark brown or black with large wings. The leaves are thick and rigid, not easily broken, sappy only when growing strongly, and with a thick epidermis.

Seed-capsules of subgenus *Robustipedunculares*

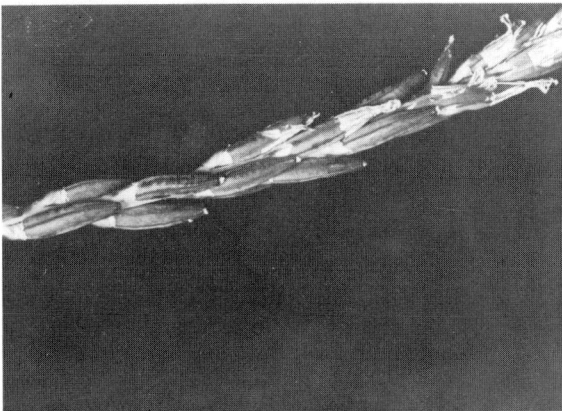

Seed-capsules of subgenus *Haworthia*

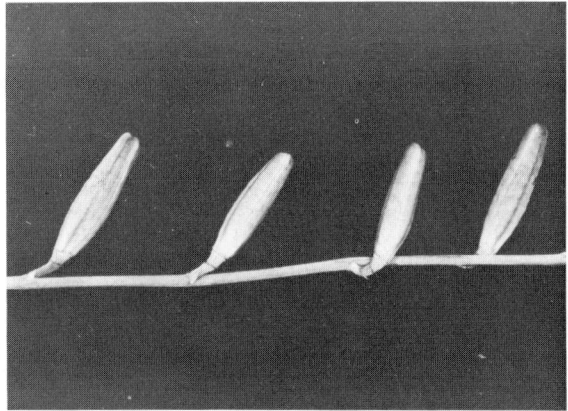

Seed capsules of subgenus *Hexangulares*

The accompanying photographs of the flowers and
fruit illustrate these differences, and for interest some
Scanning Electron Microscope photographs of seeds
of the different subgenera are included here.

All these factors help in determining relationships,
but none should be regarded in isolation.

Flowers of subgenus *Robustipedunculares*

Seed of *H. cooperi* var. *leightonii* of the subgenus
Haworthia magnified 25 times

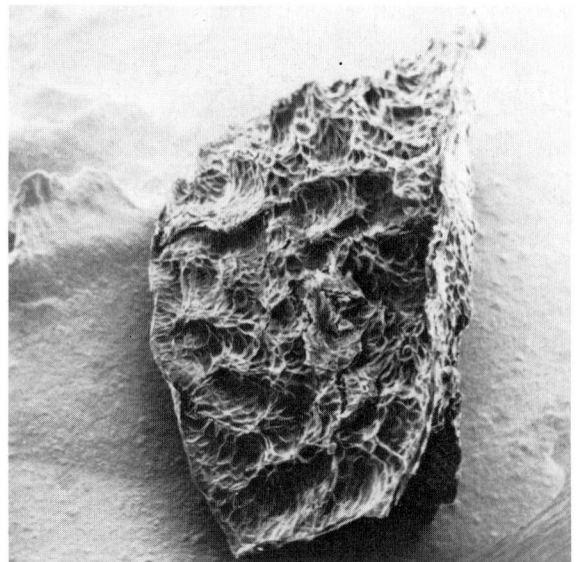

Seed of *H. retusa* var. *acuminata* of the subgenus
Haworthia magnified 25 times

Seed of *H. truncata* fa. *truncata* of the subgenus *Haworthia* magnified 15 times

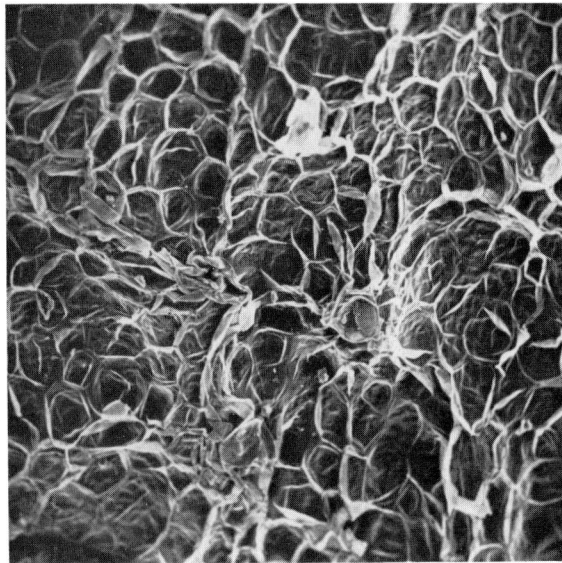

Seed of *H. truncata* fa. *truncata* of the subgenus *Haworthia* magnified 75 times

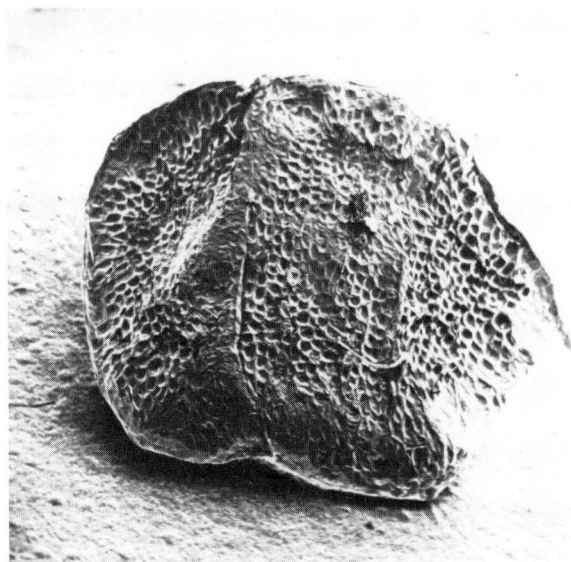

Seed of *H. xiphiophylla* of the subgenus *Haworthia* magnified 35 times

Seed of *H. fasciata* var. *fasciata* of the subgenus *Hexangulares* magnified 15 times

Seed of *H. reinwardtii* fa. *kaffirdriftensis* of the subgenus *Hexangulares* magnified 20 times

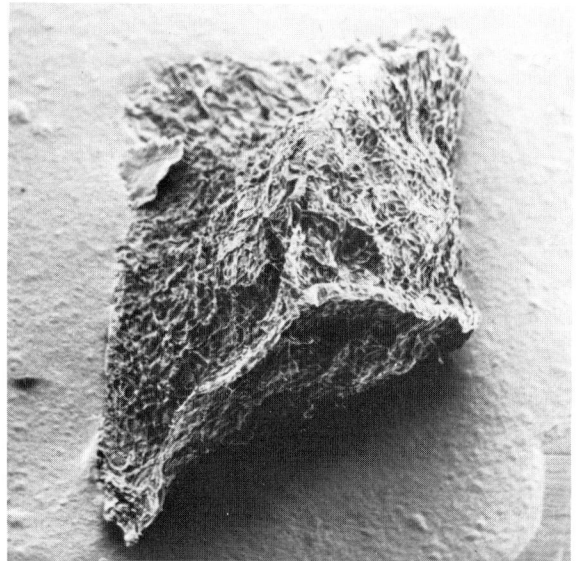

Seed of *H. scabra* var. *scabra* of the subgenus *Hexangulares* magnified 25 times

Seed of *H. viscosa* fa. *viscosa* of the subgenus *Hexangulares* magnified 20 times

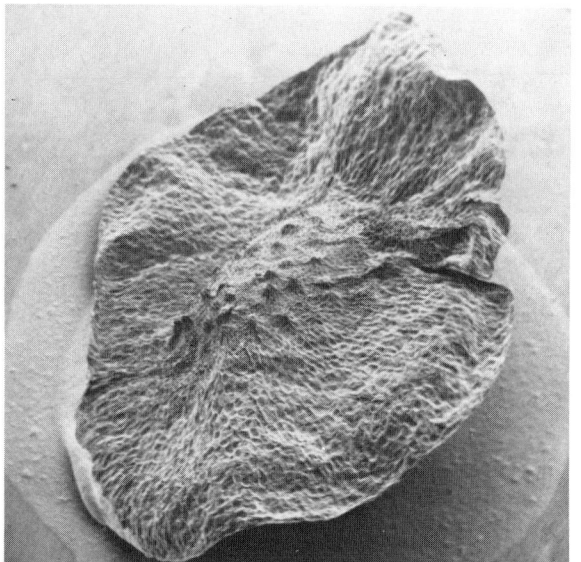

Seed of *H. minima* of the subgenus *Robustipedunculares* magnified 15 times

DISTRIBUTION OF HAWORTHIA SPECIES

Haworthias are exclusive to South Africa, with two minor exceptions. These are *H. venosa* subsp. *tessellata*, which also occurs in the south of South West Africa, and *H. limifolia*, which occurs in Swaziland and Mozambique. Apart from these two (and *H. koelmaniorum*—see below) Haworthias are confined to the Cape Province of the Republic of South Africa and to the Transkei. The greatest number of species are found between latitudes 33° and 34° south with only a few of the 'retuse' groups extending further south in the south-western Cape.

The key to understanding the distribution of the species, and the most obvious key to the delimitation of species lies in a knowledge of the main mountain ranges and river systems—see maps.

H. limifolia is a geographical oddity in the genus, which is reported at its most southerly near Stanger in Natal. There are also unconfirmed reports of this species at Kranskop and Nkandhla just beyond the midlands of Natal. It is known commonly in the drainage systems of the Pongola and Usuthu rivers and has been collected near Paulpietersburg 160km from the eastern coast. Its most northern limit appears to be at Barberton in the southern drainage area of the Crocodile river, which links with the Limpopo. Although widespread, this species is not well known and appears to occur at restricted habitats on the outskirts of its wide distribution area. It is probably common along the Ubombo mountains and is certainly plentiful in the Umbuluzi river gorge in eastern Swaziland. Records of its occurrence in Mozambique are rare. *H. koelmaniorum* is a very closely related species, if not conspecific, which was comparatively recently discovered at Groblersdal in the Olifants river drainage area (also a Limpopo tributary). This is again nearly 160km distant from Barberton, the nearest known locality for *H. limifolia*.

H. venosa is the most widespread of all the species. It is known from the Lower Breede river valley in the south-western Cape to the Orange river including the south of South West Africa, and the southern Orange Free State. It is found at 2000m above sea level at Jamestown in the north-eastern Cape to the same height above sea level in the Koue Bokkeveld mountains of the western Cape. It is essentially a species of higher lying areas and occurs on the hot,

northern slopes, but is subject to heavy frosts in the winter. Only two other species of the subgenus *Hexangulares* have comparable distribution ranges. They are *H. nigra*, which extends from near Grahamstown in the eastern Cape to Cradock in the north and then westward across the Great Karoo to northern Bushmanland; *H. viscosa* has a similar range but more southerly, extending into the Little Karoo and into the Gamtoos valley.

In the subgenus *Haworthia*, *H. arachnoidea* has the widest distribution from the north-western Cape to Namaqualand, Bushmanland, the Ceres Karoo, Little Karoo and Robertson Karoo. The delimitation of this species becomes obscure along its eastern limit, which is roughly that of the border of the summer rainfall area to about the 23rd meridian of longitude. *H. marumiana* is an unusual species in its distribution in the high mountain areas of the Great Karoo. However it is little known in the field, and there are only three confirmed reports of this species, one at Cradock, one in the Winterberg mountains south of Tarkastad and one in the Nuweveld mountains near Beaufort West. The variability of this species is quite unknown and it is not improbable that *H. batesiana*, the bright green glabrous species from the Valley of Desolation near Graaff Reinet, is continuous with it.

Other species of the Great Karoo include *H. semiviva* in the Beaufort West area, *H. bolusii* at Graaff Reinet and Pearston, moving to *H. bolusii* var. *blackbeardiana* in the north-eastern Cape from Middelburg eastwards to Queenstown and Sterkstroom. *H. cooperi* is the southern counterpart of *H. bolusii* var. *blackbeardiana* and extends into the Transkei and then to the coast at East London and westwards into the Gamtoos valley. *H. cymbiformis* is the most eastern of the Cape species, occurring from the Bashee river in the Transkei, westwards as far as Prince Alfred's Pass between Knysna and Uniondale. It extends northward to Fort Beaufort on the 33rd parallel. The only other species occurring north of this latitude is *H. nortieri*, which is associated with the Doorn and Olifants rivers in southern Namaqualand. This species has several forms varying from the narrow-leaved mountain forms on the Giftberg mountains at Van-Rhynsdorp to the larger-leaved, highly spotted forms (var. *globosiflora*) at Botterkloof.

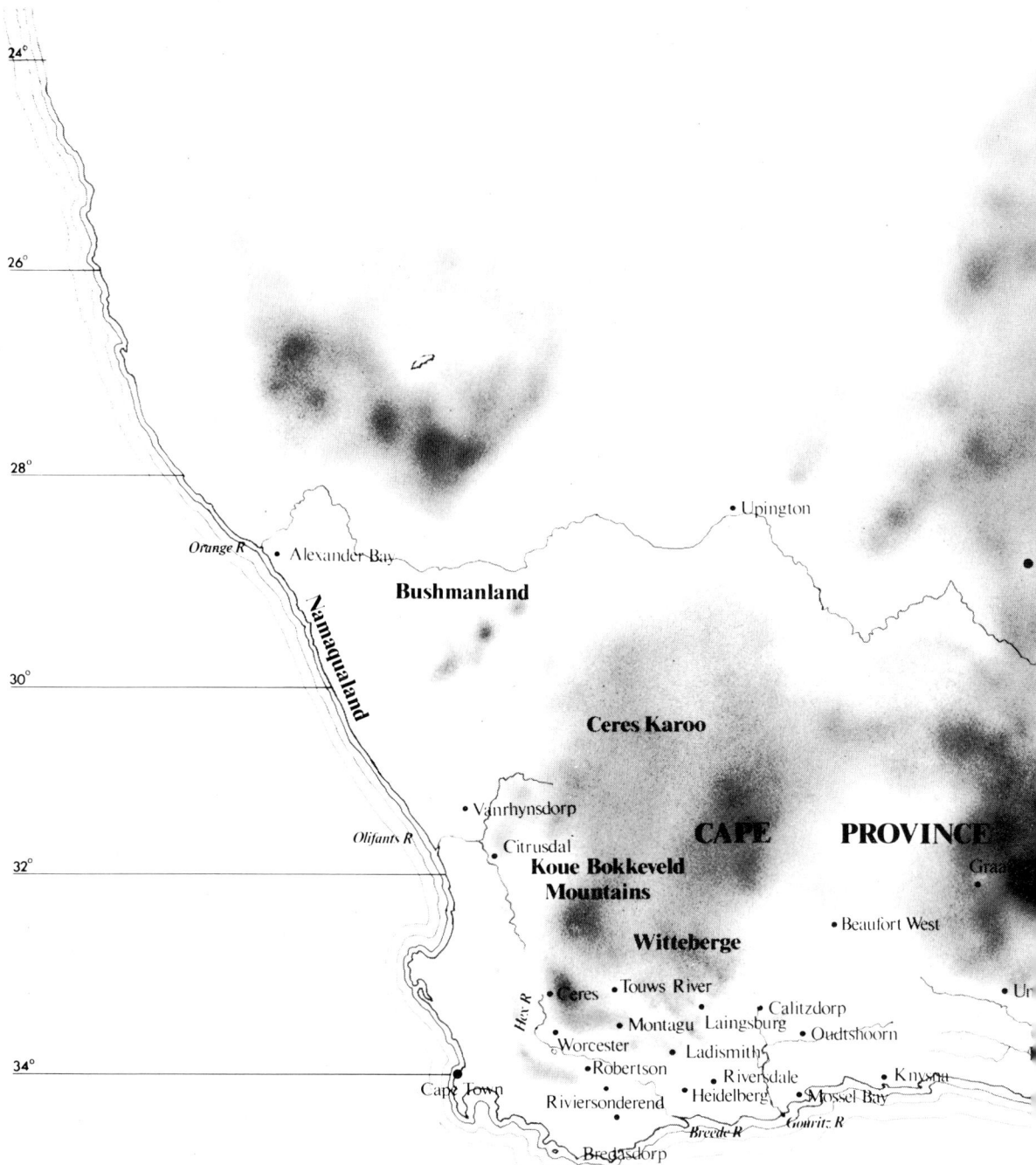

24°

26°

28°

Orange R • Alexander Bay

Bushmanland

• Upington

Namaqualand

30°

Ceres Karoo

• Vanrhynsdorp

Olifants R

• Citrusdal

CAPE   PROVINCE

32°

Koue Bokkeveld
Mountains

• Graa

• Beaufort West

Witteberge

• Ceres   • Touws River   • Ur

Hex R   • Calitzdorp

• Montagu Laingsburg   • Oudtshoorn

• Worcester   • Ladismith

34°   • Robertson   • Riversdale   • Knysna

Cape Town   • Heidelberg • Mossel Bay

• Riviersonderend   *Breede R* *Gouritz R*

• Bredasdorp

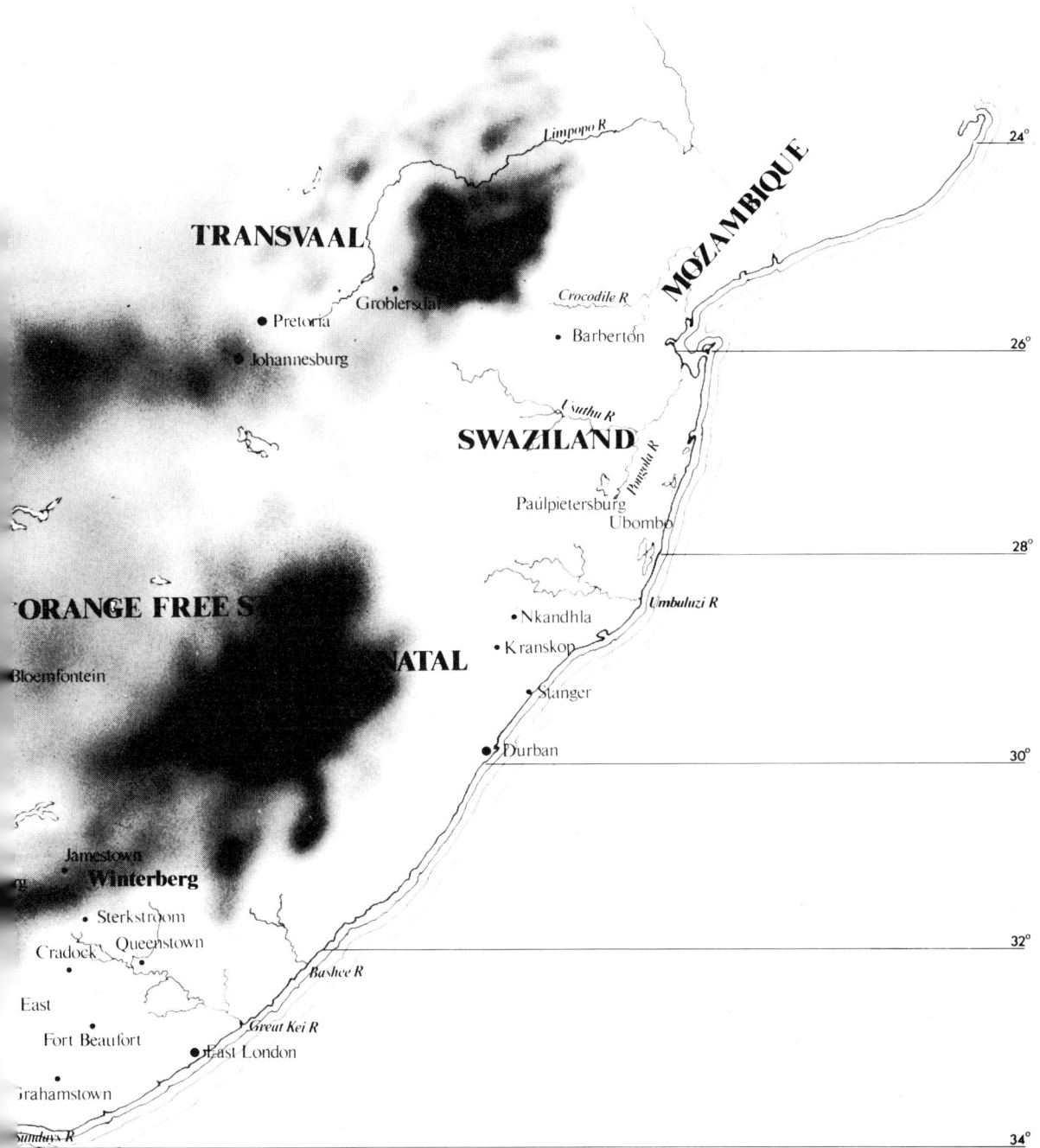

Map of southern Africa, showing mountain ranges, river systems and place names

Distribution of Subgenus Haworthia

Section Arachnoideae

Section Loratae

Section Retusae

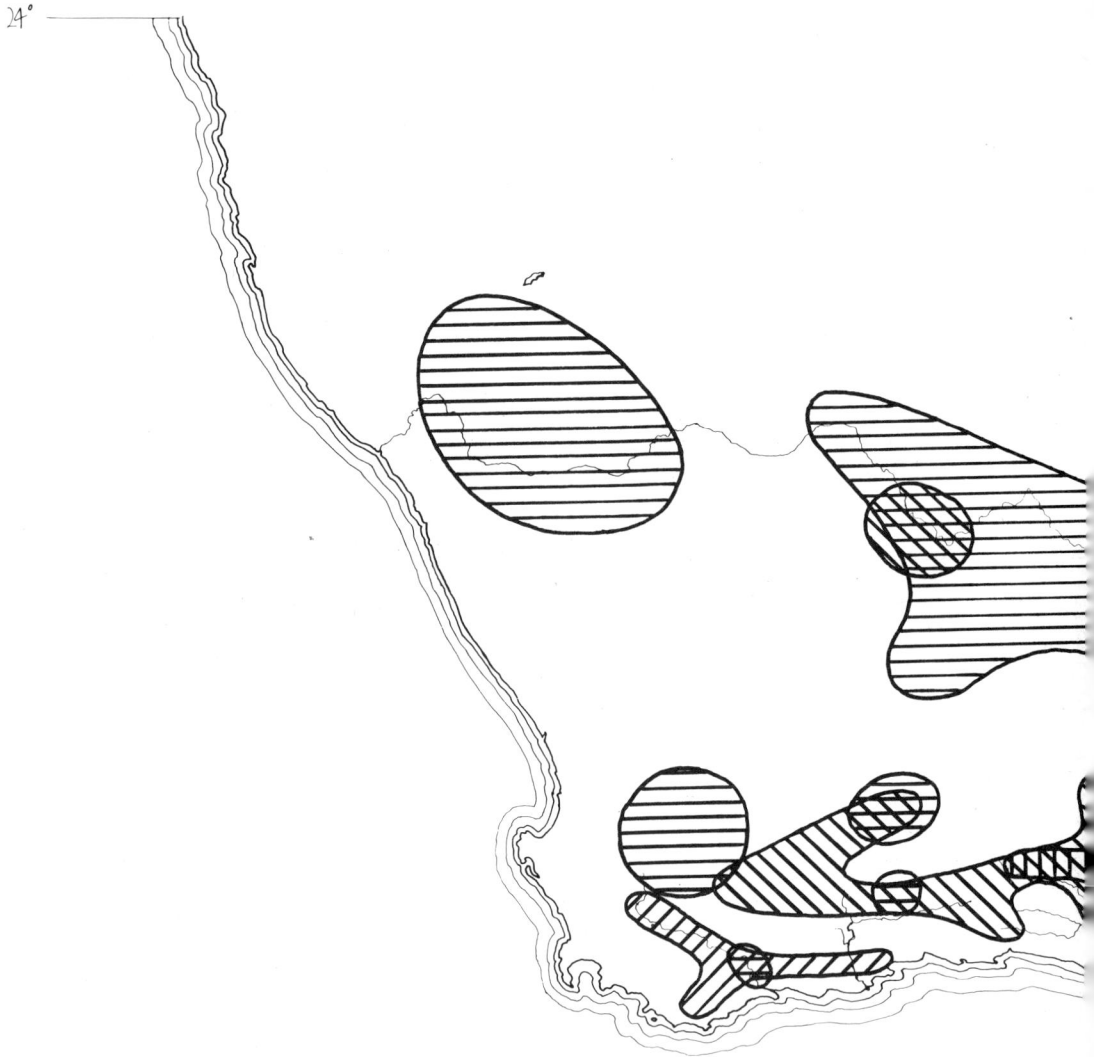

Distribution of Subgenus Hexangulares and Subgenus Robustipedunculares

Section Coarctatae

Section Trifariae

Section Venosae

Subgenus Robustipedunculares

There is a record of a comparable species at Elandskloof near Citrusdal, which needs confirmation. The relationship of this species with *H. arachnoidea* is uncertain, and the two species are not known to occur together. However, *H. arachnoidea* is known from Namaqualand and there have, so far, been no difficulties in separating the species.

Starting from the west there emerges a clear relationship between the delimitation of the species in relationship to the mountain chains and rivers. In the south-western Cape, Haworthias are first encountered south of the Riviersonderend mountains at Greyton and Genadendal, where *H. mirabilis* is found extending only as far as Stormsvlei and then southwards to Napier and Bredasdorp. *H. mutica* is also in this area but a little to the east, extending roughly along a line from the mouth of the Breede river to Riviersonderend. Both these species are represented, each by an isolated population, north of the Riviersonderend mountains at Drew in the Robertson Karoo. Other Haworthias in this western area are *H. reticulata* (which is here confused with *H. turgida*), *H. magnifica* var. *maraisii* at Stormsvlei and in isolated populations to the south, *H. pumila* only at Stormsvlei, *H. minima* along the coast and *H. marginata* at Bredasdorp. In the Riviersonderend mountains which rise from east to west, are found aberrant forms of *H. herbacea* near Villiersdorp and *H. turgida* near Greyton (the latter very closely suggests a turgid, soft form of *H. mirabilis* in the same way as *H. turgida* in the Langeberg mountains at Riversdale is related to *H. retusa*). There is also an aberrant form of presumably *H. turgida* or *H. magnifica* var. *maraisii* on the Potberg mountain at the Breede river mouth.

North of the Riviersonderend mountains is the main Breede river valley bounded on the north by the Langeberg mountains. There is the enclave of the Hex river valley opening at Worcester, and the larger enclave of the Cogmanskloof valley (Montagu) opening at Ashton. The species in this, the upper Breede river valley, or Robertson Karoo area include *H. herbacea*, *H. reticulata*, *H. pubescens*, *H. maculata*, *H. arachnoidea*, *H. magnifica* var. *maraisii*, *H. mutica*, *H. pumila*, *H. marginata* and *H. poellnitziana*.

H. pumila has an interesting distribution in that it also occurs in the upper Hex river valley, again in the Montagu enclave, while strangely enough it crosses the mountain barrier north of this and is found at one place in the Touws River valley of the Gouritz river systems. It is also recorded in the Barrydale enclave—a wholly isolated enclave of the Breede river system. *H. magnifica* is not confined to the Robertson Karoo and extends towards Heidelberg and Riversdale, also occurs at Barrydale, and recently an extraordinary record has been made of what appears to be the same species from near Laingsburg.

North of the Langeberg mountains in the west there are very few Haworthias until one reaches the Ceres Karoo where *H. arachnoidea* and *H. venosa* occur.

However there is one notable species at Touws River, *H. pulchella*, which is confined to the low mountains to the south of the town.

It is apparent that there are three latitudinal zones bounded by the Riviersonderend mountains and the Langeberg mountains. These ranges run in a west-north-westerly to east-south-easterly direction so that the Riviersonderend mountains peter out at the south of the Breede river. The Langeberg mountains link up with the Outeniqua mountains around George and Knysna. Thus moving east of the Breede river we have the southern zone now bounded by the Langeberg mountains, then the next zone which is the Little Karoo bounded by the Witteberg, Kleinswartberg and Swartberg mountains to the north, and the Great Karoo north of these mountains. In this southern zone we find firstly *H. venosa* along the lower Breede river, *H. marginata*, *H. minima* and *H. turgida* meeting with *H. reticulata*. At the Tradouw Pass are found *H. turgida* and *H. retusa* for the first time, while *H. serrata* occurs on the higher lying ridges to the south. It is at Heidelberg where the species proliferate again. *H. venosa* is absent but *H. turgida* transforms from the high sandstone form to the retuse-like shale forms. *H. retusa*, *H. floribunda*, *H. magnifica*, *H. heidelbergensis*, *H. marginata* and *H. minima* all occur. Eastwards towards Riversdale there are the same species with *H. magnifica* var. *atrofusca* and *H. magnifica* var. *paradoxa*, as well as *H. variegata*. The Gouritz river marks another break in continuity, and from Albertinia *H. chloracantha* occurs with *H. kingiana*, *H. pygmaea* and *H. parksiana* east of the Gouritz. *H. turgida* is still common here.

The situation in the Little Karoo is a little complex for two reasons. Firstly there are several internal mountain barriers, such as the Anysberg and Warmwaterberg mountains in the west, the Rooiberg in the triangle between Ladismith, Calitzdorp and Vanwyksdorp, and then the Kamanassie mountains in the east which partially plug the eastern Little Karoo off from the Great Karoo as well as from the drainage systems of both the Gamtoos and Sundays rivers. Secondly, there is the Gouritz river, which bisects the Little Karoo from north to south through Calitzdorp. J. P. Acock's well-known vegetation maps show four distinct vegetation zones in the Little Karoo and to some small extent Haworthia species are related to these. Roughly the Little Karoo can be broken into an easterly and westerly half following the Gouritz river bisection. The westerly half contains *H. viscosa*, which is not plentiful here, *H. arachnoidea* in the north-west, *H. aristata* and *H. arachnoidea* in the Barrydale enclave, *H. habdomadis* var. *inconfluens* in the central area, *H. habdomadis* var. *habdomadis* in the sandstones of the Kleinswartberg, *H. arachnoidea* at Ladismith and Amalienstein, *H. blackburniae* common in the Rooiberg area to the west of Ladismith, some *H. scabra*, and some *H. emelyae*. There is *H. magnifica* var. *major* at Muiskraal. *H. venosa* subsp. *tessellata* is scarce here and generally only found to the north west. *H. turgida* is known along the Gouritz but records are sparse. *H. wittebergensis* occurs in the mountains north of the Little Karoo and is allied both with this area and with Laingsburg in the southern Great Karoo where *H. viscosa*, *H. lockwoodii* and *H. arachnoidea* occur. The eastern section is much richer with *H. scabra*, *H. viscosa*, *H. emelyae* and *H. blackburniae* abundant. *H. habdomadis* var. *morrisiae* occurs on the red Enon conglomerates, *H. maughanii* and *H. truncata* are found in the succulent Karoo vegetation. *H. graminifolia* only occurs north of Oudtshoorn where there is also *H. starkiana*, *H. starkiana* var. *lateganiae* and *H. scabra* var. *morrisiae*. Various forms of *H. arachnoidea* occur preparatory perhaps to the change-over to *H. aranea*.

Moving a step eastwards takes us into the area of the Knysna rain forests in the south. A Haworthia has been doubtfully reported at the Storms river bridge and this needs confirmation. North of this transitional rain forest area is the still rather mountainous area between Oudtshoorn and Uniondale. Here are found *H. divergens*, *H. aristata*, *H. emelyae* and *H. aranea*, with *H. cymbiformis* var. *transiens* in the Prince Albert Pass on the boundary at the east. The area is not well known and may still yield surprises. Further north is the Great Karoo and *H. decipiens* may be encountered for the first time, with *H. nigra*, *H. venosa* subsp. *tessellata* and *H. viscosa*.

The next step eastwards is again a critical one because there are two river systems running diagonally to the sea in an east-south-easterly direction. The first is the Gamtoos system comprising the Couga, Baviaans and Groot rivers. The first two rivers drain from high rainfall sandstone areas, while the latter drains from the dry Great Karoo, and so the Gamtoos system thus has two areas for consideration. Firstly the Baviaans and Couga systems lie in the south where there are *H. cymbiformis*, *H. translucens* and *H. cooperi*, and it is not always possible to separate all the variants. Where the two rivers join the Groot river and the Karoid valley bushveld vegetation begins, occur *H. attenuata* and *H. radula*. *H. longiana* is in the conglomerate rocks, and *H. fasciata* in the higher-lying Macchia vegetation. The upper Groot River valley drains from Willowmore and Steytlerville, and this area contains *H. comptoniana*, *H. decipiens*, *H. springbokvlakensis*, *H. nigra*, *H. venosa* subsp. *tessellata*, *H. glauca* var. *herrei*, *H. woolleyi*, *H. zantnerana*, *H. sordida* and the recently discovered *H. bruynsii*.

From Port Elizabeth eastwards there is perhaps strictly no north-to-south zonation, although the Zuurberg mountains do run from the Klein Winterhoek mountains through eastwards to the Great Fish river. On the Zuurberg mountains one finds *H. glauca* var. *glauca*, *H. angustifolia* and *H. angustifolia* fa. *baylissii*, a little-known variety. *H. attenuata* and *H. fasciata* still occur and the former continues to the Kei river valley. *H. coarctata* begins at Patterson near Port

Elizabeth and *H. reinwardtii* east of Grahamstown where *H. coarctata* ends. *H. cymbiformis*, *H. cooperi* and *H. angustifolia* become common, as does *H. nigra* in the drier areas. *H. xiphiophylla* is found in the Sundays river area where *H. sordida* again occurs.

There are many problems still to be solved or better understood in the genus *Haworthia*. They include the complex of the Gamtoos river system, involving *H. translucens*, *H. cymbiformis* and *H. cooperi*. *H. viscosa* in this same area is variable and there has been a suggestion of a discrete element in the Couga valley. *H. translucens* is very variable and needs to be studied from south of Uniondale right through to Grahamstown. The occurrence of *H. attenuata* in the Kei river valley needs confirmation, as well as that of a comparable element collected in 1926 near Umtata. *H. marumiana* in the central Great Karoo and *H. batesiana* are still little known. *H. arachnoidea* is a variable species and its relationships with forms at Ladismith and at Meiringspoort or Tradouws Pass requires study. The occurrence of *H. turgida* in the Gouritz valley north of the Langeberg mountains needs investigation. *H. cooperi* is widespread and there is unconfirmed evidence of a transition in the mountains north of Bedford. Similarly *H. cymbiformis* may undergo transition with *H. batesiana* at Klipplaat west of Cathcart.

Rainfall distribution is quite interesting in relation to Haworthia species' distribution, as it is clear that many species are strictly winter growers in the wild. The south-west Cape has a Mediterranean-type climate with winter rains and for the most part annual precipitation is under 250mm per annum. The Little Karoo has rainfall peaks coincident with the equinoxes in March and September, with a hot, dry period in January and February as for the south-west Cape. The coastal strip from Swellendam to Port Elizabeth has rain at all seasons and the species which grow east of the Gouritz river appear to tolerate if not require summer watering.

This discussion has dealt with the distribution of the species in broad terms and account has to be taken of the variation introduced by smaller geographic considerations. The sub-humid to semi-arid areas in which most Haworthia species are found are vegetatively uniform. The soils are often skeletal, varying in composition over very small distances, while major geographical differences may also occur. These factors, which can play a role in the differentiation and development of species, always needs to be considered. Very often there is no problem, and in a small area four clearly different species may be found growing contiguously, but in different ecological zones. The same zones in another area may have some (very seldom all) of the same species, but now having undergone a degree of change. Still another area may present other facets of the same species and so a vast network of forms can eventually be constructed—so much so, that in one area two species may be quite distinct, yet transforming through different areas to one point where it is not possible to decide which one species occurs there. In the retuse Haworthias this situation undoubtedly occurs with the complication that six or more elements may be involved in the mosaic at any one time. It follows therefore that attempting to name plants in collections can often prove impossible with precision without some idea of the origin of the plants concerned.

CHECKLIST OF SPECIES

This list consists of those species and lower taxa which are accepted in this book. The detailed commentary on them is in the chapter 'Commentary on Species'.

Haworthia altilinea
  angustifolia fa. *angustifolia*
  angustifolia fa. *baylissii*
  angustifolia fa. *grandis*
  angustifolia fa. *paucifolia*

  arachnoidea

  aranea

  archeri var. *archeri*
  archeri var. *dimorpha*

  aristata var. *aristata*
  aristata var. *helmiae*

  armstrongii

  attenuata fa. *attenuata*
  attenuata fa. *britteniana*
  attenuata fa. *caespitosa*
  attenuata fa. *clariperla*

  batesiana

  blackburniae

  bolusii var. *bolusii*
  bolusii var. *blackbeardiana*

  bruynsii

  chloracantha var. *chloracantha*
  chloracantha var. *denticulifera*
  chloracantha var. *subglauca*

  coarctata subsp. *coarctata* var. *coarctata* fa. *coarctata*
  coarctata subsp. *coarctata* var. *coarctata* fa. *conspicua*
  coarctata subsp. *coarctata* var. *coarctata* fa. *chalwinii*
  coarctata subsp. *coarctata* var. *tenuis*
  coarctata subsp. *adelaidensis* fa. *adelaidensis*
  coarctata subsp. *adelaidensis* fa. *bellula*

  comptoniana fa. *comptoniana*
  comptoniana fa. *major*

  cooperi var. *cooperi* fa. *cooperi*
  cooperi var. *cooperi* fa. *pilifera*

  cooperi var. *leightonii*

  cymbiformis var. *cymbiformis* fa. *cymbiformis*
  cymbiformis var. *cymbiformis* fa. *gracilidelineata*
  cymbiformis var. *cymbiformis* fa. *multifolia*
  cymbiformis var. *cymbiformis* fa. *obesa*
  cymbiformis var. *cymbiformis* fa. *planifolia*
  cymbiformis var. *cymbiformis* fa. *ramosa*
  cymbiformis var. *incurvula*
  cymbiformis var. *transiens*
  cymbiformis var. *umbraticola*

  decipiens

  divergens

  emelyae var. *emelyae*
  emelyae var. *multifolia*

  fasciata fa. *fasciata*
  fasciata fa. *browniana*
  fasciata fa. *sparsa*

  floribunda

  glabrata

  glauca var. *glauca*
  glauca var. *herrei* fa. *herrei*
  glauca var. *herrei* fa. *jacobseniana*
  glauca var. *herrei* fa. *jonesiae*

  graminifolia

  habdomadis var. *habdomadis*
  habdomadis var. *inconfluens*
  habdomadis var. *morrisiae*

  heidelbergensis

  herbacea

  cv. *kewensis*

  kingiana

  koelmaniorum

  cv. *kuentzii*

  limifolia var. *limifolia* fa. *limifolia*
  limifolia var. *limifolia* fa. *major*
  limifolia var. *gigantea*
  limifolia var. *striata*

lockwoodii

longiana

maculata

magnifica var. *magnifica*
magnifica var. *atrofusca*
magnifica var. *major*
magnifica var. *maraisii*
magnifica var. *meiringii*
magnifica var. *notabilis*
magnifica var. *paradoxa*

cv. *mantelii*

marginata

marumiana

maughanii

minima

mirabilis subsp. *mirabilis* fa. *mirabilis*
mirabilis subsp. *mirabilis* fa. *beukmannii*
mirabilis subsp. *mirabilis* fa. *napierensis*
mirabilis subsp. *mirabilis* fa. *rubrodentata*
mirabilis subsp. *mirabilis* fa. *sublineata*
mirabilis subsp. *badia*
mirabilis subsp. *mundula*

mutica

nigra fa. *nigra*
nigra fa. *angustata*
nigra fa. *nana*

nortieri var. *nortieri*
nortieri var. *globosiflora*

cv. *ollasonii*

parksiana

poellnitziana

pubescens

pulchella

pumila

pygmaea fa. *pygmaea*
pygmaea fa. *crystallina*
pygmaea fa. *major*

radula

reinwardtii var. *reinwardtii* fa. *reinwardtii*
reinwardtii var. *reinwardtii* fa. *chalumnensis*
reinwardtii var. *reinwardtii* fa. *kaffirdriftensis*
reinwardtii var. *reinwardtii* fa. *olivacea*
reinwardtii var. *reinwardtii* fa. *zebrina*
reinwardtii var. *brevicula*

reticulata var. *reticulata*
reticulata var. *hurlingii*
reticulata var. *subregularis*

retusa var. *retusa* fa. *retusa*
retusa var. *retusa* fa. *fouchei*
retusa var. *retusa* fa. *geraldii*
retusa var. *retusa* fa. *longebracteata*
retusa var. *retusa* fa. *multilineata*
retusa var. *acuminata*
retusa var. *dekenahii*

rycroftiana

scabra var. *scabra*
scabra var. *morrisiae*

semiviva

serrata

sordida

springbokvlakensis

starkiana var. *starkiana*
starkiana var. *lateganiae*

cv. *subattenuata*

translucens subsp. *translucens*
translucens subsp. *tenera*

truncata fa. *truncata*
truncata fa. *crassa*
truncata fa. *tenuis*

turgida fa. *turgida*
turgida fa. *caespitosa*
turgida fa. *pallidifolia*
turgida fa. *suberecta*

ubomboensis

variegata

venosa subsp. *venosa*
venosa subsp. *granulata*
venosa subsp. *tessellata*

viscosa fa. *viscosa*
viscosa fa. *asperiuscula*
viscosa fa. *beanii*
viscosa fa. *pseudotortuosa*
viscosa fa. *subobtusa*
viscosa fa. *torquata*

wittebergensis

woolleyi

xiphiophylla

zantnerana.

COMMENTARY
ON
SPECIES

COMMENTARY ON SPECIES

Listed alphabetically below are all those species, subspecies, varieties, forms and cultivars recognized, and, for convenience, those listed in Jacobsen's *Handbook of Succulent Plants* (English edition, 1960, Volume 2) or his *Succulent Lexicon* (English edition, 1974), which have been discarded since those books went to print, or which are not acknowledged herein, with their appropriate synonymy.

H. aegrota von Poelln., *Des. Pl. Life* 11:193 (1939)
Synonymous with *H. herbacea*.

H. altilinea Haw., *Phil. Mag.* 44:301 (1824); *Haw. Handb.* 96 (1976), Bayer; *Nat. Cact. Succ. Journ.* 34:53 (1979), Scott
Section Arachnoideae subsection Limpidae
Although Bayer regards this species in his Handbook as confused, having been inconsistently applied over the years, there are in cultivation plants which accord with the excellent illustration in Salm-Dyck's monograph on Aloes and Mesembryanthemaceae (captioned *Aloe altilinea* 11 fig. 3, and *Aloe altilinea denticulata* 11 fig 3B). They are of easy cultivation, strongly growing and clustering freely. Rosettes are about 8 to 10cm or more across fully grown and stemless. Leaves are dark green, varying according to the amount of light they are given, broadly lanceolate, with more or less prominent teeth on the margins and the upper part of the keel.

Scott's application of this name, as set out in the 1979 article in the National Cactus & Succulent Society's journal (ref. above), is confused. His fig. 1 is of a plate in the Herbarium Library at the Royal Botanic Gardens, Kew, which, when examined, confirms its more appropriate placing near *H. cooperi*, the colouring particularly confirming this. Neither this painting nor the plants figured at figs 3 and 4 of the article are the same as the aforementioned illustration (reproduced in his article as fig. 2), which is taken as typifying the species.

Reported from the Uitenhage Division at Redhouse and near the Zwartkops river; Stockenstrom Division on stony ridges near Seymour; Grahamstown; Cathcart; Zwartkops mountains in the Prince Albert District; Hankey near Port Elizabeth; Longrove Plantation near Port Elizabeth. These localities reported in the 1930s to von Poellnitz.

H. angustifolia Haw., *Phil. Mag.* 46:283 (1825)
Section Loratae subsection Loratae
This is a dull green, opaque-leaved, small-rosette species with toothed leaf margins and keel. Haworth's original description was brief, and von Poellnitz expanded it considerably in 1936 (*Fedde's Repert. Spec. Nov.* 40:149 (1936)): rosette stemless, remaining solitary or sparingly suckering from the base (in cultivation nowadays it seems to cluster readily), with rosettes 5 to 6cm in diameter with about 20 leaves, the

H. altilinea

H. angustifolia fa. *angustifolia*

H. angustifolia fa. *baylissii*

H. angustifolia fa. *grandis*

leaves lanceolate, tapering, 4 to 5cm long and 1cm broad at the base, young leaves erect, older leaves recurved, not very flexible, the tip with a whitish end-bristle, upper surface concave at the base, flat above, often with a broad, prominent, concolourous central line and 5 to 7 indistinct longitudinal lines beneath the surface. The surface of the leaves is matt and rough from a covering of minute tubercles, visible only with aid. Several varieties are listed by Jacobsen in his Handbook and Lexicon, contributed by various authors, but their standing as varieties is not upheld by Bayer after field study of this species and the neighbouring closely related *H. chloracantha*. He disposes of these varieties as follows:

var. *albanensis* (Schönl.) von Poelln., *Rec. Albany Mus.* 2:256 (1912) and *Fedde's Repert. Spec. Nov.* 41:194 (1937), synonymous with the type.

var. *subfalcata* von Poelln., *Sukkulentenkunde* IV (1951)—erroneously referred by Bayer to *H. magnifica* var. *maraisii* (*New Haw. Handb.* 44 (1982)), merely a larger variant of the type with somewhat sideways curving leaves, not warranting separation at any level.

var. *liliputana* Uitew., *Succulenta* (1953) 43, synonymous with *H. chloracantha* var. *denticulifera*.

var. *denticulifera* von Poelln., *Fedde's Repert. Spec. Nov.* 41:194 (1937), referred as a variety of *H. chloracantha*.

He relegates Scott's *H. baylissi* to *forma* level beneath *H. angustifolia*.

The species *H. angustifolia* resolves then as follows:

H. angustifolia Haw. fa. *angustifolia* (syn. *H. albanensis* Schönl., *H. angustifolia* var. *albanensis* (Schönl.) von Poelln., *H. angustifolia* var. *subfalcata* von Poelln.);

H. angustifolia fa. *baylissii* (Scott) Bayer, *New Haw. Handb.* 26 (1982);

H. angustifolia fa. *grandis* (G. G. Smith) Pilbeam stat. nov. (syn. var. *grandis* G. G. Smith, *Journ. S.A. Bot.* 9: 105 (1943));

H. angustifolia fa. *paucifolia* (G. G. Smith) Pilbeam stat. nov. (syn. var. *paucifolia* G. G. Smith, *Journ. S.A. Bot.* 14: 48 (1948))

The description of the type is as given above, but the variability and the sinking of the former varieties must be borne in mind. Descriptions of the other forms are:

fa. *baylissii*, with broad leaves more spreading than others in the species, except perhaps the discredited var. *subfalcata*; reported from the farm Oudekraal near the Zuurberg pass;

fa. *grandis*, rosette stemless, slowly proliferous from the base, up to 10cm tall, leaves about 30, the young ascending, incurved near the tip, the older ascending and spreading, 6 to 10cm long and recurved, gradually narrowed from the base to the tip, light yellowish-green, towards the tips of the old leaves becoming reddish, not shining; from Albany Division, collected on a rocky hill facing north, fully exposed to the sun;

fa. *paucifolia*, rosette stemless, not or slowly proliferous from the base, up to 6cm tall, about 2cm in diameter near the base. Leaves about 12, young ascending and recurved as they elongate, older leaves ascending, long-recurved, to 5cm, gradually tapering from near the base to the tip, very dark, dull green, margins minutely toothed. Collected near Frasers Camp, in the Grahamstown district, some miles from the type locality of fa. *grandis*, growing in grass veld in full sun. Reported from Great Brak River near

H. angustifolia fa. *paucifolia*

Avontuur; Bredasdorp; Little Brak River; Calitzdorp—these localities for the type; also from Fernkloof near Grahamstown; Boterklof near Grahamstown; Howieson Port, north-east of Grahamstown; Montagu; Calitzdorp; Riversdale; Great Brak River—these localities for the former var. *albanensis*; also from Albany Division—fa. *grandis*; Grahamstown Division, near Frasers Camp—fa. *paucifolia*; Bonnievale, Robertson Division—var. *subfalcata*.

H. arachnoidea (L.) Duv., *Pl. Succ. Hort. Alenc.* 7 (1809); *Spec. Pl.* 322 (1753), Linnaeus; *Monogr.* 12:2 (1836–63), Salm-Dyck; *Haw. Handb.* 27 (1976), Bayer; *Cact. Amer.* 49:205 (1977), Scott; *New Haw. Handb.* 27 (1982), Bayer

Section Arachnoideae subsection Arachnoideae

This is a species whose identity has for years been in doubt due to its synonymy with *H. setata*, which name has overshadowed it. But it is by priority of publication the prevailing name, and is indeed the type species for the genus. *H. setata* is reduced by Bayer in his Handbook to synonymy together with its varieties, which based as they are on size of rosette, length or density of bristles or coloration, do not warrant more than at most forma status, and for most of them not even that, except for var. *xiphiophylla*, which is reinstated as a separate species by Bayer.

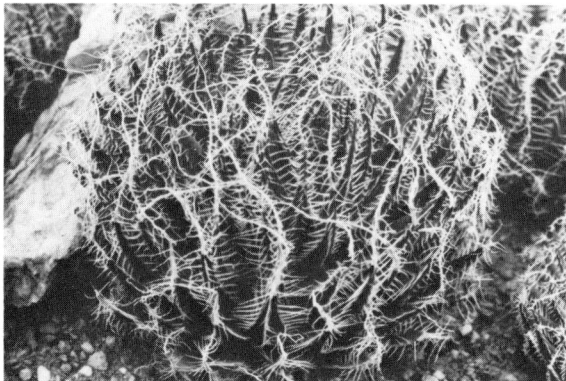

H. arachnoidea

H. arachnoidea is not a quick-growing species, and will not tolerate excessive watering, especially if water is allowed to lie among the leaves for any length of time. In time, however, it will clump, often dividing dichotomously, to form attractive low mounds of outward-facing rosettes. It requires an open compost, and even less water when resting than when growing, when watering should be modest. The characteristic, thick, densely-flowered flower-stems appear in early spring in England, and the flowers exude a great deal of nectar. The leaves differ from similar species in being opaque and uniformly green, with rarely a hint of translucence at the very tips of the leaves; the amount and density of the characteristic bristles varies

considerably. Reported from a wide area: Robertson, Wellington, Bredasdorp Division; Little Karoo; Ladismith and Oudtshoorn Division; Port Elizabeth and Uitenhage Division; the eastern part of the Great Karoo; Willowmore and Steytlerville Division; Sutherland Division; Little Namaqualand; Bushmanland. Areas cited for former varieties are: Little Namaqualand; Springbok, Richtersveld; Bonnievale; Ezelsjacht Poort; Oudtshoorn; Bredasdorp; 3 miles east of Port Nolloth; Breekpoort; Lekkersing—var. *bijliana* von Poelln.; Amalienstein near Ladismith, Little Karoo—var. *gigas* von Poelln.; Ladismith—var. *joubertii* von Poelln.; Wellington—var. *nigricans* Haw.

H. arachnoidea

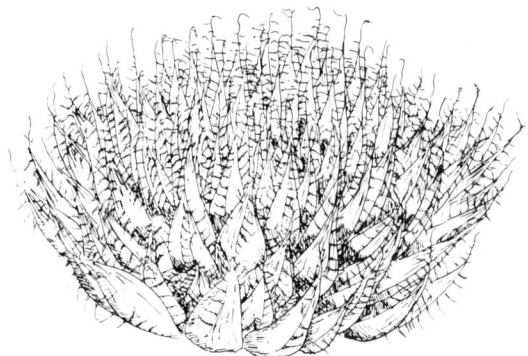

H. arachnoidea

H. aranea (Berg.) Bayer, *Haw. Handb.* 98 (1976); *Pflanz.* 33:114 (1908), Berger

Section Arachnoideae subsection Arachnoideae

Bayer in his Handbook has uplifted this former variety of *H. bolusii*, ascribed thereto by Berger in 1908, to specific status, after extensive study in the field. It makes a flatter rosette than *H. bolusii*, with a depressed centre, giving a flat-bowl shape, with narrower leaves than *H. bolusii*, finely and densely bristled, and making an altogether larger rosette. It is certainly worth a place alongside *H. bolusii* in any collection, being distinctive enough to determine from others similar in appearance, and beautiful enough to be counted among the choicer species.

It represents the same cultural problems as *H. bolusii*, in its intolerance of excessive water, especially if left wet between the leaves, when rot is liable to set into the wide stem from which the leaves arise, reducing the whole plant to a translucent, squashy mess very quickly. And with this very flat growing sort of plant, with little length of stem there is little hope of saving it once rot has developed. I suspect that in time it might form clusters, dividing dichotomously perhaps as some others in this section do, but plants I have had in cultivation for ten years have remained solitary and seem to have settled for a maximum diameter of about 13 to 15cm. It flowers in England in early spring with the characteristic close-packed flower-stem, with short pedicels and upward angled flowers coloured grey-green.

Reported from the western parts of the Great Karoo. Bayer says that 'It occurs in the Macchia vegetation between Oudtshoorn and Uniondale, particularly in the Herold/Molen River/De Rust area . . . North-east of Oudtshoorn one finds larger coarser forms which may be more directly referable to *H. arachnoidea*.'

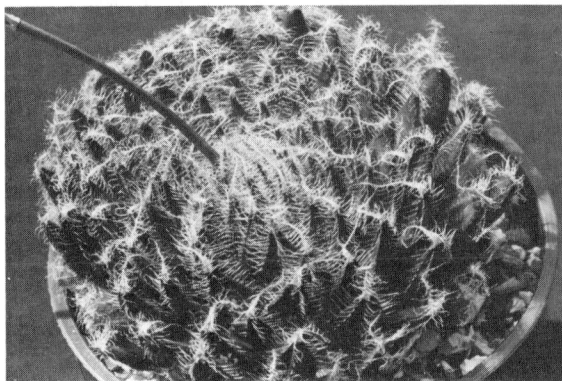

H. aranea

H. archeri Barker ex Bayer, *Journ. S. A. Bot.* 47:791 (1981); *New Haw. Handb.* 29 (1982), Bayer
Section Retusae subsection Turgidae

Very recently described and practically unknown in cultivation as yet, the relationships of this good-looking, small-growing species are unclear. Bayer believes it most closely resembles *H. marumiana*, which it certainly does, but without the proliferating habit of that species. It is a small, brownish-green, many leaved rosette, at its maximum about 6cm in diameter, with narrow-lanceolate leaves, spreading and recurving slightly to erect, inflated at the middle of the leaves, tapering to fine points, with prominent bristles or teeth on the margins and keel and on the surfaces near the tips of the leaves, and with translucent flecks numerous towards the tips.

Reported from Baviaan, west Lainsgsburg, and from Ngaap Kop, west Laingsburg.

H. archeri var. *archeri*

H. archeri var. *dimorpha*, Bayer, *Journ. S. A. Bot.* 47:793 (1981); *New Haw. Handb.* 29 (1982), Bayer, is similar to the type, but with broader, flatter leaves, more recurving than the type, and with larger, more spaced teeth, more confined to the margins, keel and a centre line on the upper surface of the leaves.

Reported from Constable Station, west Laingsburg.

H. archeri var. *dimorpha*

H. aristata Haw., *Suppl. Pl. Succ.* 51 (1819); *Cact. Amer.* 18:24 (1946), J. R. Brown; *Nat. Cact. Succ. Journ.* 35:11 (1980), Scott
Section Arachnoideae subsection Arachnoideae

Although Bayer rejected this name in his Handbook, Scott has pointed out that there was a contemporary illustration in the Herbarium Library of the Royal Botanic Gardens at Kew, which undoubtedly lends weight to the consideration of this name in preference to Bayer's choice of the later *H. unicolor*.

The species forms clumps of grass-green rosettes, with tapering, wisp-ended leaves, which will dry up quickly at the tips if allowed to get too dry. The leaves are triangular in cross-section, very turgid and soft in texture.

Reported from the western Karoo region near Barrydale, and east of Lemoenshoek, and near Ladismith, near Calitzdorp and towards Oudtshoorn, and at several localities in the Little Karoo. Without sufficient reference, Scott disposes of *H. venteri* (which Bayer maintains as a variety of *H. unicolor*) as synonymous with *H. aristata*, and certainly *H. venteri* co-incides very well with the Kew illustration of *H. aristata*.

H. aristata var. *aristata*

H. aristata var. *aristata*

H. aristata var. *helmiae* (von Poelln.) Pilbeam comb. nov. is similar to the type, but altogether more compact, with many more leaves, and tending to grow lower in the soil. It is not by any means quick-growing, making a solitary rosette, low to the ground, with a strong tendency to die back and curl the leaves in over the heart of the plant when resting. Watering should be carefully restrained at all times. The localities reported by von Poellnitz are, according to Bayer, highly confusing, and he allies the species to plants found at a site south of Cango Caves, Oudtshoorn.

H. aristata var. *helmiae*

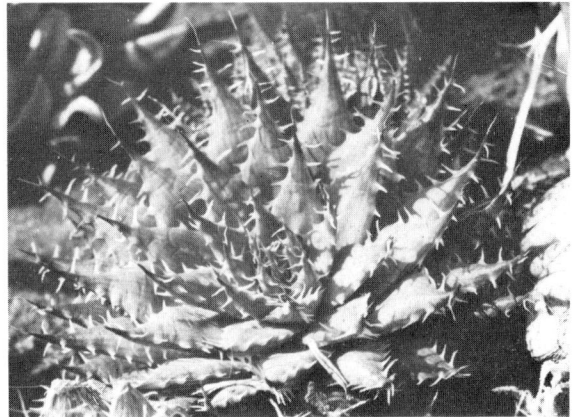

H. aristata var. *aristata* (toothed form)

H. armstrongii von Poelln., *Kakteenk.* 1937.152 (1937); *Cact. Amer.* 10:123 (1939) & 22:60 (1950), J. R. Brown; *The First Fifty Haw.* 11 (1970), Pilbeam; *Haw. Handb.* 99 (1976), Bayer; *Journ. S.A. Bot.* 47:540 (1981), Brandham & Cutler
Section Coarctatae subsection Coarctatae

Reduced by Bayer to *forma* status beneath *H. glauca* var. *herrei*, but reinstated, after cytological studies by Brandham and Cutler, to its former specific level.

This is a slender, grey-green leaved species, from only one reported locality, where it is believed to have originated from one clone, 9km north-east of Uitenhage. The leaves are triangular in outline, spear-shaped, with generally sparse rows of whitish tubercles confined to the margins and keels of the leaves. It makes columnar stems to about 15cm tall, 3 to 4cm wide, offsetting from the base and lower stem.

H. asperiuscula Haw., *Suppl. Pl. Succ.* 60 (1819); *Cact. Amer.* 25:164 (1953), J. R. Brown
Reduced to form status beneath *H. viscosa*.

H. armstrongii

H. attenuata fa. *attenuata*

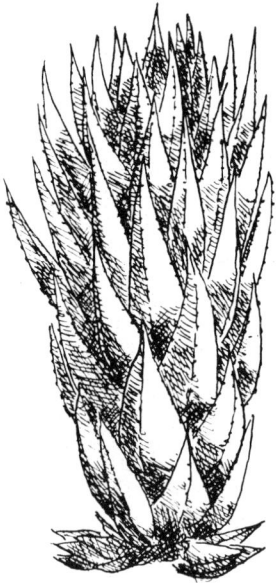

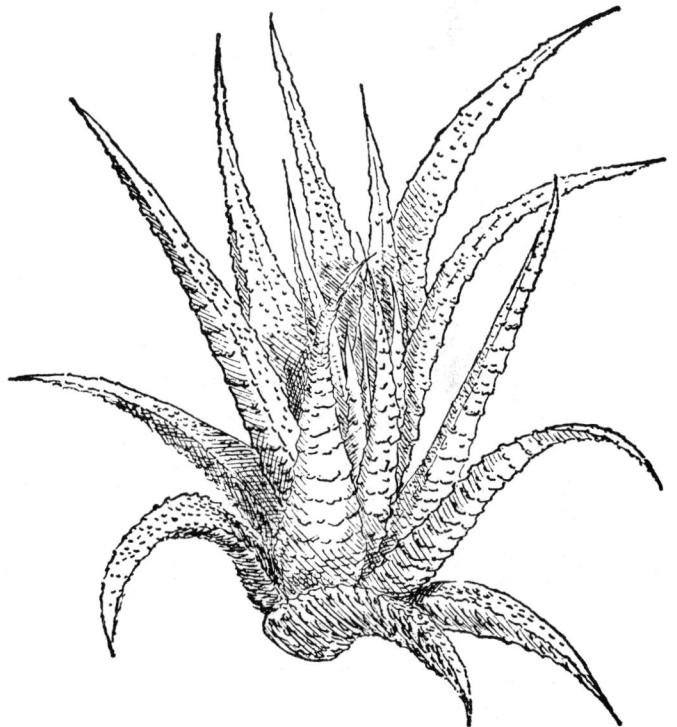

H. armstrongii

H. asperula Haw., *Phil. Mag.* 300 (1824); *Cact Amer.* 28:78 (1956), J. R. Brown; *The First Fifty Haw.* 11 (1970), Pilbeam

Bayer dismisses this name as indeterminate, and, apart from an illustration in Salm-Dyck's Monograph near to *H. emelyae*, and a tendency to apply this name to larger-growing plants of *H. pygmaea*, there has been little real idea of its application.

H. atrofusca G. G. Smith, *Journ. S.A. Bot.* 14:41 (1948); *Cact. Amer.* 25:78 (1953), J. R. Brown; *Haw. Handb.* 100 (1976), Bayer; *Nat. Cact. Succ. Journ.* 32:18 (1977), Bayer

Referred by Bayer to varietal status beneath *H. magnifica*.

H. attenuata Haw., *Syn. Pl. Succ.* 92 (1812); *The First Fifty Haw.* 11 (1970), Pilbeam; *The Second Fifty Haw.* 18 (1975), Pilbeam; *Haw. Handb.* 100 (1976), Bayer

Section Coarctatae subsection Attenuatae

This is a very variable species, common in collections under a plethora of varietal names or in despair with no varietal name at all, so indeterminate are the differences between them. Many of the varieties were erected by R. S. Farden in 1939 (*Cact. Journ.* 8:34 (1939)), but examination of his papers in the Lindley Library of the Royal Horticultural Society leads me to the clear conclusion that his work was ill-founded and the varieties he erected (many in any case invalidly under ICBN rules) should be discarded. An anomaly

H. attenuata fa. *attenuata*

which has crept by the back door into the names under this species is *H. attenuata* var. *caespitosa* (Berg.) Farden. This variety started its printed life as a variety of *H. fasciata*, understandably enough because of its markings similar to that species. But there the similarity ends, as *H. fasciata* is a columnar species with incurving, rigid leaves, while this variety of *H. attenuata* is short-stemmed with flexible, erect to recurving leaves like the type of *H. attenuata*, much more attenuated than the ovate-lanceolate leaves of *H. fasciata*. Consequently Farden quite rightly ascribed it to *H. attenuata*. Bayer erred in confusing this variety with *H. attenuata* var. *clariperla*. Apart from fa. *clariperla*, fa. *britteniana* (and fa. *caespitosa*, which is here maintained for reasons given below), the remaining forms of *H. attenuata* vary merely in the density and size of the tubercles, a characteristic not warranting recognition at any level, since the variation is very wide. These forms are the most distinctive and worthy of recognition. The species resolves as follows:

H. attenuata Haw. forming a short-stemmed rosette, with dark green, long, flexible, gradually tapering, recurving leaves, clustering from the base; leaves occasionally not recurving, but not incurving. The back surface slightly to heavily spotted with white tubercles, upper surface less so, but never free from tubercles.

H. attenuata fa. *attenuata*

H. attenuata fa. *britteniana*

H. attenuata fa. *caespitosa*

H. attenuata fa. *caespitosa* (Berg.) Pilbeam stat. nov., *Pflanz.* 4.38:92 (1908), Berger; *Cact. Journ.* 8:34 (1939), Farden; *Succulenta* 34:38 (1952), Uitewaal; *The First Fifty Haw.* 11 (1970), Pilbeam; non Bayer, *Haw. Handb.* 107 (1976); non Bayer, *New Haw. Handb.* 63 (1982). This form is so widely grown as to be unnecessary to describe, but it is often still confused with *H. fasciata*, and Bayer has now confused it with fa. *clariperla*, so that it is worth spelling out the differences. Its form is the same as the type, with a short stem and long, flexible, recurving leaves; the outstanding character and distinguishing feature of this form, apart from a more noticeable vigour, is the coalescing into transverse bands of the tubercles on the backs of the leaves, which is the cause of the confusion (which continues in most general books published) with *H. fasciata*. A most popular form of *H. attenuata*, clustering rapidly, tolerating most collectors' treatment and looking well on it, it is at its best if given a wide pan, rich compost and sufficient water in the growing period (summer) to keep it growing fast, otherwise it follows the tendency in all long-leaved species to dry up at the leaf-tips; good light will intensify the contrast of the tubercles with the leaf base-colour of shining dark green. Reported from Port Elizabeth and Fort Brown.

H. attenuata fa. *caespitosa*

H. attenuata fa. *caespitosa* (selected form with prominent banding)

H. attenuata fa. *caespitosa* (variegated form)

H. attenuata fa. *clariperla* (Haw.) Pilbeam stat. nov., *Phil. Mag.* 3:186 (1826); *Monogr.* 6:12b (1836–63), Salm-Dyck; *Journ. Linn. Soc.* 18:203 (1880), Baker; *The First Fifty Haw.* 11 (1970), Pilbeam; non Bayer, *Haw. Handb.* 107 (1976); non Bayer, *New Haw. Handb.* 63 (1982). This is a distinct form with intensely white tubercles, both large and small (but separate and not coalescing into bands to any marked degree as in fa. *caespitosa*) on a dark green background colour; it is beautifully illustrated in Salm-Dyck's Monograph and in the collection of watercolours in the Herbarium Library at Kew.

H. attenuata fa. *britteniana* (von Poelln.) Bayer, *New Haw. Handb.* 63 (1982)—incorrectly as *H. attenuata* fa. *britteniae*; *Fedde's Repert. Spec. Nov.* 41:196 (1937), von Poelln. This form has both large and small prominently white tubercles thickly on the reverse of the leaves and often quite thickly too on the upper surface. It is found in Plutos Vale.

H. attenuata fa. *clariperla*

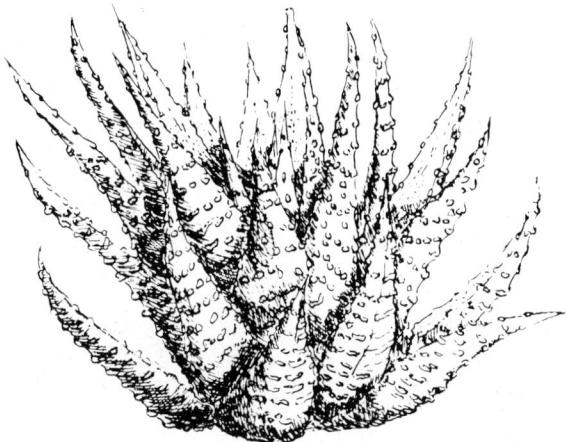

H. attenuata fa. *clariperla*

H. baccata G. G. Smith, *Journ. S.A. Bot.* 10:20 (1944); *Cact. Amer.* 38:5 (1966), J. R. Brown; *Haw. Handb.* 101 (1976), Bayer

This species is taken to be a variant within Bayer's concept of the species *H. coarctata*, as no sort of population resembling that pictured with Smith's original description could be found at the type locality of Isidenge, about 9 miles south-west of Stutterheim, and there is some confusion as to where it was collected.

H. badia von Poelln., *Kakteenk.* 76 (1938); *Aloe* 11:8 (1973), Scott; & 12:89 (1974), Bayer; *The Second Fifty Haw.* 20 (1975), Pilbeam; *Haw. Handb.* 101 (1976), Bayer

Referred by Bayer to subspecific status beneath *H. mirabilis*.

H. batesiana Uitew., *Nat. Cact. Succ. Journ.* 3:101 (1948); *Cact. Amer.* 26:52 (1954), J. R. Brown; *The First Fifty Haw.* 11 (1970), Pilbeam; *Haw. Handb.* 101 (1976), Bayer; *New Haw. Handb.* 30 (1982), Bayer
Section Arachnoideae subsection Proliferae

H. batesiana

This species is popular and widespread in collections, due to its ease of cultivation and clustering habit, which lead to wide distribution. It has shining green leaves, duller in strong light conditions, subtly tessellated beneath the surface with dark veining, tapering from the turgid centre of the leaf to a long, whispy tip. It forms dense cushions of small rosettes quickly, each rosette only about 4 or 5cm wide, and it has one of the most attractive flowers in this small-flowered genus: larger lobed than most, short stemmed (about 16cm overall), with glistening, snowy-white petals. It was described by the Dutch collector and student of this genus, A. J. A. Uitewaal, in honour of the notable collector of the pre-war years in England, J. T. Bates. It is found in the Valley of Desolation near Graaff Reinet, and is probably widespread around this area. Bayer suggests that it may be a smooth form of *H. marumiana*.

H. batesiana

H. batteniae C. L. Scott, *Cact. Amer.* 51: 268 (1979)
   This is clearly a redescription of *H. blackbeardiana*.

H. baylissii C. L. Scott, *Journ. S.A. Bot.* 34:1 (1968);
Haw. Handb. 102 (1976), Bayer; *New Haw. Handb.* 26
(1982), Bayer.
   Referred to *H. angustifolia* at form level.

H. beanii G. G. Smith, *Journ. S.A. Bot.* 10:137
(1944); *Cact. Amer.* 31:150 (1959) & 32:188 (1960), J.
R. Brown; *Haw. Handb.* 102 (1976), Bayer
   Bayer rejects this species and the smaller variety, *H. beanii* var. *minor*, as probably synonymous with *H. viscosa*, which occurs in the same area. Plants in cultivation are generally much more slow-growing than other forms of *H. viscosa*. Because of its distinct appearance it is maintained at form level beneath *H. viscosa*.

H. bilineata Bak., *Journ. Linn. Soc.* 18:213 (1880)
   This is an uncertain species and it is difficult, if not impossible, to determine what was being referred to in the original description. No habitat details of any use were given, no herbarium specimens deposited, and no photographs or drawings appeared at the time, or have appeared since with any real certainty as to their identity. Karl von Poellnitz evidently thought he knew what constituted the species, because he reduced Baker's *H. affinis*, itself considered now as indeterminate, to varietal status beneath it, and for good measure erected another variety, *H. bilineata* var. *gracilidelineata*. A contemporary drawing of the latter by J. T. Bates (dated 1933) is of a plant commonly in cultivation, a very translucently leaved form of *H. cymbiformis*, near *H. cymbiformis* var. *incurvula*, maintained at form level under that species. There is in cultivation a plant answering pretty well to Baker's description, with its obligatory two lines at the upper margins of the leaves, and it appears to be a form of *H. cymbiformis*, but with so little real idea of the originally described plant, it is best discarded as insufficiently known.

H. bilineata

H. blackbeardiana von Poelln., *Fedde's Repert. Spec. Nov.* 31:82 (1932) & 41:196 (1937) & 44:236 (1938); *Des. Pl. Life* 9:9 (1937); *Cact. Amer.* 22:4 & 76 (1950), J. R. Brown; *The Second Fifty Haw.* 21 (1975), Pilbeam; *Haw. Handb.* 103 (1976), Bayer; *New Haw. Handb.* 31 (1982), Bayer
   Referred now by Bayer to varietal status beneath *H. bolusii*.

H. blackburniae Barker, *Journ. S.A. Bot.* 3:93 (1937); *Kakteenk.* 37 (1938), von Poellnitz; *Cact. Amer.* 15:14 (1943), J. R. Brown; *The Second Fifty Haw.* 21 (1975), Pilbeam; *Haw. Handb.* 103 (1976), Bayer
Section Loratae subsection Fusiformes

H. blackburniae

   This is an extreme of the genus in its habit, with its swollen, tuber-like roots and attenuated, stiff, grass-like foliage. For the former of these characteristics it was placed in a separate section by Miss Barker, when she described the species, 'fusiformes' meaning spindle-shaped, and referring to the swollen roots. It was rapidly followed by the species *H. graminifolia*, and would perhaps have been joined by *H. wittebergensis*, had not this latter discovery been found not to have tuberous roots, although in all other characteristics it belongs here. The leaves of *H. blackburniae* are variable in length, width and thickness: up to 20cm long, up to 5mm wide and only 1mm thick, but varying from plant to plant, some having noticeably more slender leaves with less rigidity than others whose leaves retain their shape through more rigidity.

It is found in the Oudtshoorn Division 'in stony ground on a hill about 8 miles from Calitzdorp in quite exposed conditions'. The history of its discovery bears repetition: when it was found no real idea of its status generically could be gained, since none collected was in flower, and it was only after Mrs Blackburn, the discoverer of the species in 1935, had sent plants in flower to G. W. Reynolds of Johannesburg, that it was realized that this new discovery was a Haworthia. In fact Reynolds had expressed the opinion, before seeing the flowers, that it was a new species of Leptaloe. After seeing the flowers both he and Miss Verdoorn were of the opinion that Haworthia was the correct genus for this new discovery, and the new section was set up.

H. blackburniae

H. blackburniae (short-leaved form)

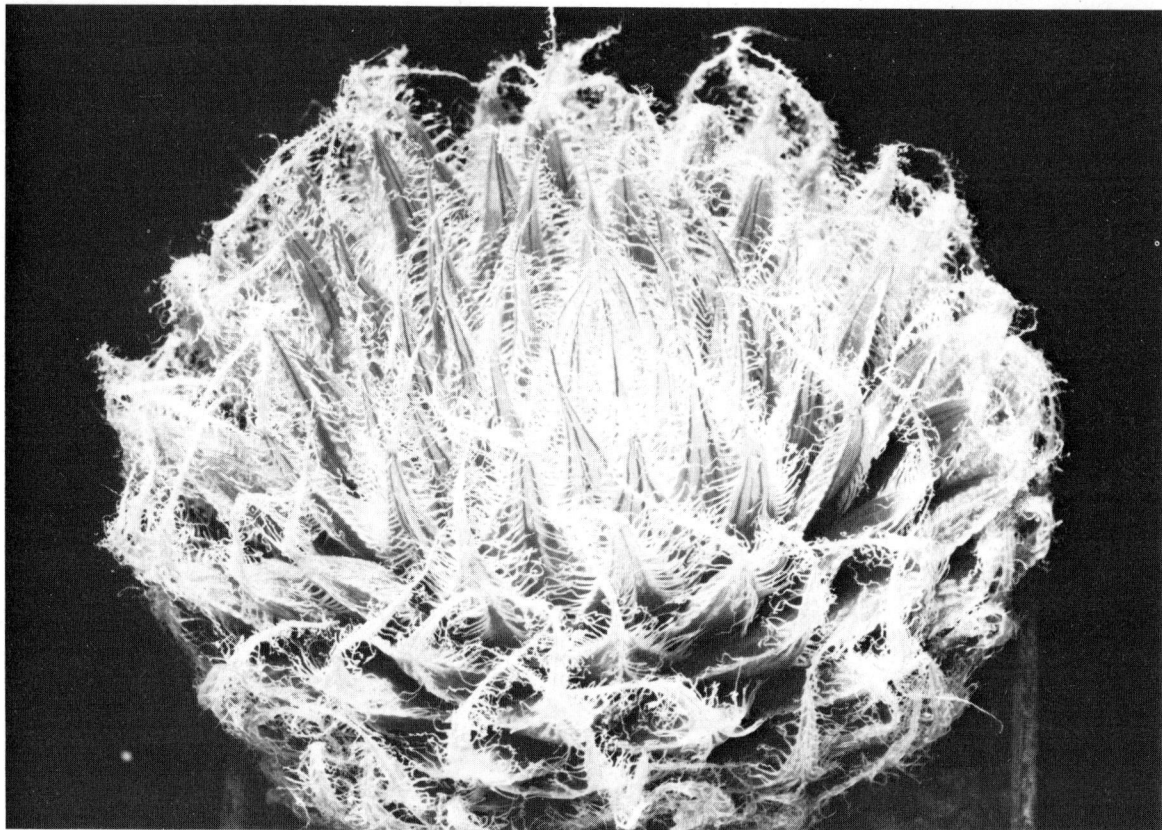

H. bolusii var. *bolusii*

Bayer reports that it is common in quartzitic rocks between the Rooiberg Pass south of Vanwyksdorp and extends to about 20 miles west of Ladismith; it generally prefers a cooler, protected, southern aspect, where it forms dense clumps. Although it does not present undue problems in cultivation once established, it is no easy task to get collected plants going, as almost invariably the roots seem to have been left behind in the soil where they were collected. But once rooted it will grow as readily as most other touchy species, given care with exposure to too much sunshine, and a gentle hand with the watering-can. It grows readily and quite quickly from seed.

This name was used, just after its application here by Miss Barker, by von Poellnitz for another species. He subsequently changed the name appropriately to *H. correcta*.

H. bolusii Bak., *Journ. Linn. Soc.* 18:215 (1880); *Fedde's Repert. Spec. Nov.* 27:134 (1929–30) & 44:135 (1938), von Poellnitz; *Cact. Amer.* 13:87 (1941) & 17:34 (1945), J. R. Brown; *The Second Fifty Haw.* 24 (1975), Pilbeam; *Haw. Handb.* 104 (1976), Bayer; *New Haw. Handb.* 31 (1982), Bayer
Section Arachnoideae subsection Limpidae

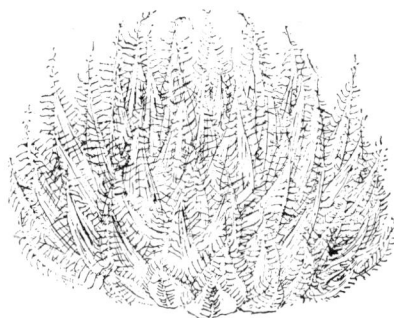

H. bolusii var. *bolusii*

This species, second only perhaps to the truncate species, did much to popularize this genus among growers of succulent plants in the 1950s and 1960s. Its multitudinous, fine bristles on bluish-green or greyish-green leaves, with large areas of translucence at the tips, glisten in the light and make it a desirable acquisition in any collection. Its only disadvantage from a collector's point of view is that its propagation must be from seed, since it does not readily cluster, although it does sometimes divide dichotomously or will throw up an adventitious offset on a flower-stem. In addition, to add to the collector's woes it has the habit of rotting quickly and irretrievably if water is allowed to remain in the rosette for any length of time, especially in damp, cold weather. Rosettes vary somewhat in size, from about 5cm in diameter to the larger variety, newly transferred by Bayer, var. *blackbeardiana*, which is anything up to 15cm in diameter.

H. bolusii var. *blackbeardiana* (von Poelln.) Bayer, *New. Haw. Handb.* 31 (1982), Bayer, has been reduced from its former species status, since Bayer maintains that it cannot be readily separated. It is a much more vigorous variety, presenting few of the difficulties of the type, although its wide stem will not tolerate overwatering at the wrong time of year.

H. bolusii var. *blackbeardiana*

Reported from a wide area from Middelburg to Sterkstroom, southwards to Graaff Reinet and Cradock, south-west in the eastern Karoo to Klipplaat and Jansenville, and north-west near Barclay East, the latter smaller with fewer but larger leaves.

Former varieties (var. *aranea* and var. *semiviva*) have been restored to species level by Bayer.

H. bolusii var. *blackbeardiana*

H. browniana von Poelln., *Des. Pl. Life* 9:102 (1937); *Fedde's Repert. Spec. Nov.* 44:212 (1938); *Cact. Amer.* 18:163 (1946), J. R. Brown; *The Second Fifty Haw.* 25 (1975), Pilbeam; *Haw. Handb.* 105 (1976), Bayer

This species has been reduced to form status under *H. fasciata* by Bayer. Although this is believed to be, like *H. armstrongii*, from only one clone, Brandham has indicated (in letters) that the origins could be from *H. fasciata*, the chromosome mutation it carries causing a change in morphology. Its treatment as a form of *H. fasciata* is therefore acceptable.

H. bruynsii Bayer, *Journ. S.A. Bot.* 47:789 (1981); *New Haw. Handb.* 64 (1982), Bayer

A most extraordinary recent discovery, this species confounds the writer's concept of classification (and everyone else's) by having all the appearance of being related to species in the Section Retusae (it grows near the similar *H. springbokvlakensis*) but producing hexangular flowers! For the moment therefore it is not allocated to a Section, until its origins are properly determined.

It makes a small rosette, keeping low to the ground, with recurved leaves, with distinctly cut-off end-areas, these areas rounded-triangular and roughened from small papillae, translucent and with somewhat acute margins. It grows well down in the soil beneath bushes at the two known localities south-east of Steytlerville.

H. bruynsii

H. caespitosa von Poelln., *Fedde's Repert. Spec. Nov.* 41:197 (1937) & 44:232 (1938); *Cact. Journ.* 6:18 (1937); *Des. Pl. Life* 9:90 (1937); *Haw. Handb.* 105 (1976), Bayer

This species is referred to *H. turgida* by Bayer; it is maintained here at form level under that species.

H. carrissoi Res., *Bol. Soc. Brot.* 15:161 (1941); *Haw. Handb.* 106 & 177 (1976) Bayer
  Referred by Bayer to *H. glauca*.

H. cassytha Bak., *Fl. Cap.* 6:337 (1896); Haw Handb. 106 (1976), Bayer

H. cassytha

H. coarctata subsp. *adelaidensis* fa. *adelaidensis*

Dismissed by Bayer, plants seen under this name have generally been plants somewhere near *H. tortuosa*, itself in Bayer's opinion probably of hybrid origin.

H. chloracantha Haw., *Revis. Pl. Succ.* 57 (1821); *Haw. Handb.* 106 (1976), Bayer
Section Loratae subsection Loratae

  Although this species has been known for many years it is by no means common in collections, perhaps because it has not many of the attractions of some Haworthia species more popular with the general collector. It is a uniform, dull green, with leaves triangular in section, toothed at the margins and keel. Rosettes are about 6cm in diameter, clustering from the base. There is an excellent early illustration of this species in Salm-Dyck's Monograph (51:1). There has been confusion between this species and *H. angust-ifolia*, and indeed the two species are closely related. But field studies indicate that the former small varieties of the latter, typified by var. *liliputana* and var. *denticulifera* (and Bayer does not uphold both names) are more correctly ascribed to *H. chloracantha*, which species name Bayer upholds for the whole complex from the Mossel Bay to Great Brak area,

H. chloracantha var. *chloracantha*

H. chloracantha var. *subglauca*

saying that 'there are no easy distinctions to make between this species and *H. angustifolia* in the East Cape except that it has more robust widely spaced marginal teeth which tend to be concolorous or often darker in colour than the leaves'. The type, *H. chloracantha* var. *chloracantha*, is the robust, proliferous, light green form found on steep, north-facing, conglomerate slopes, north of Herbertsdale. *H. chloracantha* var. *subglauca* von Poelln. is retained for the glaucous-green, also robust, but less proliferous form on the granite soils around Great Brak. *H. chloracantha* var. *denticulifera* (von Poelln.) Bayer is

the small, moderately proliferous, dark green variety found in the vicinity of Mossel Bay. The species resolves as follows:

H. chloracantha Haw. var. *chloracantha*.

H. chloracantha var. *denticulifera* (von Poelln.) Bayer (syn. *H. angustifolia* var. *denticulifera* von Poelln. and var. *liliputana* Uitew.), *Fedde's Repert. Spec. Nov.* 41:194 (1937) von Poelln.; *Succulenta* 43 (1953), Uitewaal; *Haw. Handb.* 112 (1976), Bayer; *H. chloracantha* var. *subglauca* von Poelln., *Kakteenk.* 135 (1937).

Cultivation presents no real problems, but watering should not be too liberal. A light position is to be preferred to keep the leaves compact and well coloured, especially in the case of var. *subglauca*, which will turn quite blue, showing up the prominent darker teeth on the margins of the leaves well. The type and var. *denticulifera* tend to redden in full light.

H. chloracantha var. *denticulifera*

H. chloracantha var. *denticulifera*

H. coarctata Haw., *Phil. Mag.* 44:301 (1824); *Nat. Cact. Succ. Journ.* 28:80 (1973), Bayer; *Haw. Handb.* 107 (1976), Bayer
Section Coarctatae subsection Coarctatae

This species, along with *H. reinwardtii*, was the subject of a paper by Bayer (1973 reference above), which made some sense of the mass of names which had accumulated like flotsam around these two species. Under *H. coarctata* Bayer put *H. greenii* and some former varieties of *H. reinwardtii*, so that the species became reclassified at that time as follows:

H. coarctata Haw. subspecies *coarctata* var. *coarctata*;

H. coarctata subspecies *coarctata* var. *greenii* (Bak.) Bayer;

H. coarctata subspecies *coarctata* var. *tenuis* (G. G. Smith) Bayer;

H. coarctata subspecies *adelaidensis* (von Poelln.) Bayer.

The basis on which the revision was made was complicated and made difficult reading, but to the Haworthia-attuned eye it made good sense on unravelling, and once the basic concept is accepted other basic differences become apparent, like the generally shorter, rounder-backed leaves of *H. coarctata* as reconstituted, the less prominent, more concolorous tubercles, the more sharply uncurving leaf-tips and the subtle differences in coloration compared with *H. reinwardtii*, especially in plants grown in sunny conditions.

Former varieties worth retaining at form level because of their distinct appearance and value to collectors as distinct forms are: (former *H. reinwardtii* varieties) var. *chalwinii*, var. *conspicua*, var. *bellula*; but var. *greenii* is not upheld here, since it varies continuously with *H. coarctata* in habitat, as confirmed in letters by Bayer, who has expressed doubts about the validity of its varietal status since his placing it there in his Handbook.

H. coarctata subsp. *adelaidensis* fa. *bellula*

H. coarctata is one of the most easily cultivated species, and if room is given in a wide container without too much depth being necessary it will form a large clump within a few years, with the exception of the very small form, *bellula*, which is extremely slow. I have seen photographs of this species in habitat carpeting the ground so thickly that it was really impossible to walk in the area without treading on the plants. I like to see cultivated plants with a good deal of colour, which can be achieved by exposure to sunlight for about half of the day in England and generous watering in the summer months, with enough water at any time to stop the compost drying out completely. There is a tendency for this and similar species to dry up the leaf-tips if left too dry for too long, or even for whole leaves to dry up in an unsightly way. This should not occur with plants receiving sufficient water, but if it does the offending leaves may be winkled out with a pair of tweezers, with care, for fear of decapitating the whole stem.

The species, with retained forms, resolves as follows:

H. coarctata Haw. subsp. *coarctata* var. *coarctata* fa. *coarctata*;

H. coarctata subsp. *coarctata* var. *coarctata* fa. *chalwinii* (Marl. et Berg.) Pilbeam comb. nov. (syn. *H. chalwinii* Marl. et Berg. in *Notizbl. Bot. Gart. Mus. Berl.* 4:247 (1906) & *H. reinwardtii* var. *chalwinii* (Marl. et Berg.) Res., *Mem. Soc. Brot.* 67 (1943));

H. coarctata subsp. *coarctata* var. *coarctata* fa. *conspicua* (von Poelln.) Pilbeam comb. nov. (syn. *H. reinwardtii* var. *conspicua* von Poelln. in *Fedde's Repert. Spec. Nov.* 41:210 (1937));

H. coarctata subsp. *coarctata* var. *tenuis* (G. G. Smith) Bayer, *Nat. Cact. Succ. Journ.* 28:80 (1973); *Journ. S.A. Bot.* 14:51 (1948), G. G. Smith;

H. coarctata subsp. *adelaidensis* (von Poelln.) Bayer, *Nat. Cact. Succ. Journ.* 28:86 (1973); *Beitr. Sukk.* 2:43 (1940), von Poelln.;

H. coarctata subsp. *adelaidensis* fa. *bellula* (G. G. Smith) Pilbeam comb. nov. (syn. *H. reinwardtii* var. *bellula* G. G . Smith, *Journ. S.A. Bot.* 11:70 (1945)).

They are distinguished as follows:

var. *coarctata* fa. *coarctata* (the type) covers the wide variation in this species, but generally is taken to refer particularly to less heavily tubercled, larger forms of the species. The variation in marking is considerable, from almost smooth (former var. *greenii*) to distinctly spotted leaves, from long to short and from wide to narrow leaves, but constantly shining green, colouring red in full light, and forming dense clumps of columnar, leaf-covered stems;

fa. *chalwinii* is well-known in collections, and was described as long ago as 1906. It has short, thick, wide leaves in the shape of the hull of an old galleon, heavily and regularly spotted. A well-grown clump will grace any collection or show-bench, especially if given sufficient light to colour the leaves red-brown and so intensify the whitish coloured tubercles; it is found along the Kowie River;

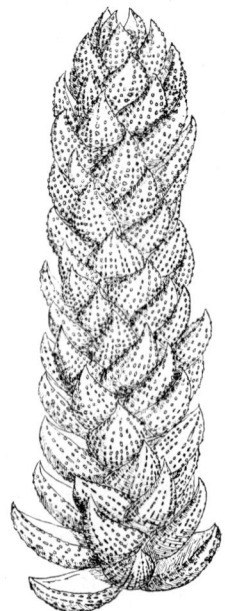

H. coarctata subsp. *coarctata* var. *coarctata* fa. *chalwinii*

H. coarctata subsp. *coarctata* var. *coarctata* fa. *coarctata* (all these forms occurring in the same area at Howiesonspoort)

H. coarctata subsp. *coarctata* var. *coarctata* fa. *coarctata*

H. coarctata subsp. *coarctata* var. *coarctata* fa. *chalwinii*

H. coarctata subsp. *coarctata* var. *coarctata* fa.
conspicua

fa. *conspicua* is a really large-growing form with
large, wide leaves incurving strongly and patterned in a
distinctive, crazy way with thinnish, broken bands of
tubercles wandering across the backs of the leaves. It
was collected between Port Elizabeth and Alexandria;

var. *tenuis* is a popular variety in cultivation, as it is
prolific and easily grown, lending itself to wide
distribution. The long, rope-like stems flop over under
their own weight when they get to 12 or 13cm long, and
trail down to root where they touch the soil; a good
candidate to fill a large pan quickly;

H. coarctata subsp. *coarctata* var. *tenuis*

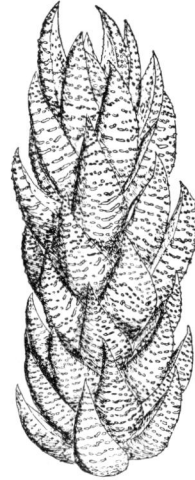

H. coarctata subsp. *coarctata* var. *coarctata* fa.
conspicua

subsp. *adelaidensis* is a very distinctive, beautiful
subspecies, which incidentally does not occur as one
might expect at Adelaide: von Poellnitz received
specimens from W. E. Armstrong, who lived at
Adelaide, and as was so often the case, confused the
habitat of the plants with the habitat of the sender. The
type comes from Queens Road, north-east of Grahamstown, an easterly population of the subspecies in
higher areas. It is perhaps the most attractively marked
of the species, with narrow, loosely arranged leaves
with prominently-white tubercles in regular rows on
the backs of the leaves. It is not such a robust grower as
others of the species, with stems rarely more than
about 13cm tall and 3cm wide, and it takes some time to
produce a small cluster;

fa. *bellula* is really one of the most charming forms of
the whole species, with stems rarely more than 5cm tall
and clustering to form low clumps of petite appearance. It came from a restricted locality '4½ miles
from Grahamstown on the Cradock Road'. Perhaps its
diminutive stature has come about from harsh growing
conditions; certainly it is one of the slowest and most
difficult to grow well in cultivation. Bayer in his 1982
Handbook indicates that this form may have been
collected to extinction, since he found none on visiting
the type locality. G. G. Smith, when describing it
originally, referred to it as being found by several
collectors, so that it clearly occurred in some numbers.

H. coarctatoides Res. & Viveiros, *Port. Acta Biol.*
2:176 (1948)
Referred to *H. coarctata*.

H. comptoniana fa. *comptoniana*

H. comptoniana fa. *major*

H. comptoniana G. G. Smith, *Journ. S.A. Bot.* 11:76 (1945); *Cact. Amer.* 34:182 (1962), J. R. Brown; *Aloe* 11:8 (1973), C. L. Scott; *Aloe* 12:89 (1974), Bayer; *The Second Fifty Haw.* 28 (1975), Pilbeam
Section Retusae subsection Retusae

This is a beautiful, glossy-surfaced member of this section of beautiful species, with white flecks and 5 to 7 prominent, white lines, tessellately connected beneath the surface in the almost flat, translucent end-area. It forms a rosette of 8 to 9cm in diameter with about a dozen leaves, reluctantly clustering from the base in time, although leaves may be rooted to form new plants. The leaves are recurved at a right-angle to form a flat top to the rosette, and are thick and turgid, 4 to 5cm long, 2cm broad at the base of the triangular end-area and about 2.5cm long in this part, with a small end-bristle at the tip of the leaf, where the leaf often incurves slightly. It is indeed a distinctive, striking species worth seeking out. Bayer has suggested that it varies continuously with *H. emelyae*, but has not yet thought fit to merge them. In view of these species' distinct appearance I am happy to maintain them separately here. G. G. Smith reported it from the Willowmore Division; Bayer reports a larger growing form from the Georgida area west of Willowmore, which is described below at form level. It is twice the size of the type, more vigorous and very handsome.

It is not difficult to grow, although rapidity of growth is not its strong feature. An open compost is

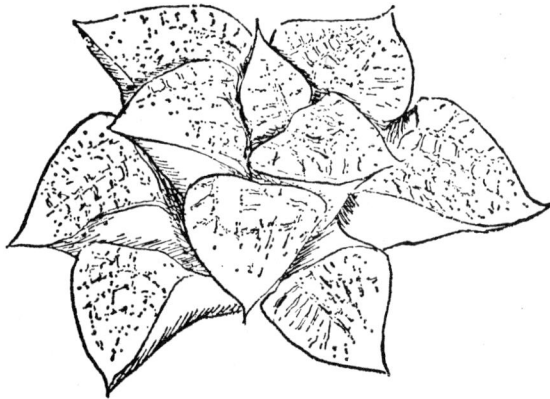

H. comptoniana fa. *comptoniana*

recommended with up to 50 per cent grit.

H. comptoniana fa. *major*, forma nov., *a typo foliis majoribus differt*. Differs from the type in having larger leaves.

H. cooperi Bak., *Saund. Refug. Bot.* 4.t.233 (1870); *Cact. Amer.* 39:46 (1967), J. R. Brown & 46:170 (1974), Bayer & Pilbeam; *Haw. Handb.* 109 (1976), Bayer; *The First Fifty Haw.* 15 (1970), Pilbeam
Section Arachnoideae subsection Limpidae

In a paper in the American Society's journal (ref. above) the author in collaboration with Bayer rein-

stated the species *H. pilifera* from its aberrant relegation to varietal status under *H. obtusa*. At the time of this paper other parts of this variable species needed further field study, and the hint was made with regard to *H. cooperi* and *H. vittata* that these species were fairly certainly northern ecotypes of *H. pilifera*, representing introgression of *H. pilifera* towards *H. blackbeardiana*, and that the whole group might well result in one variable species. Bayer subsequently in his Handbook took the name *H. cooperi* as preceding *H. pilifera* (although he maintained *H. blackbeardiana* and subsequently in his *New Haworthia Handbook* allied it to *H. bolusii*), and did not maintain *H. pilifera* at any level. In view of the considerable difference in appearance of extreme forms of this widespread species, the name is maintained at form level for the blunter, more round-tipped, generally slower-growing forms. *H. cooperi* var. *leightonii* is maintained by Bayer for the former species *H. leightonii* (NOT 'leightoniae'—a persistent mis-spelling, which would mean incidentally that Captain Leighton after whom the plant was named had changed his sex), for distinctive plants found a few miles inland from Kaysers Beach, west of East London; its distinguish-

H. cooperi var. *cooperi* fa. *cooperi*

H. cooperi var. *cooperi* fa. *cooperi*

H. cooperi var. *cooperi* fa. *cooperi*

ing character is the reddish coloration of the veins beneath the leaf-surface in the translucent end-area of the grey-green leaves; it occurs, Bayer reports, 'in shallow, sandy soils and partially covering high, granite slabs'. The species resolves as follows:

H. cooperi Bak. var. *cooperi* fa. *cooperi*;

H. cooperi var. *cooperi* fa. *pilifera* (Bak.) Pilbeam stat. nov. (syn. *H. pilifera* Bak., *Saund. Refug. Bot.* 4.t.234 (1870));

H. cooperi var. *leightonii* (G. G. Smith) Bayer (syn. *H. cooperi* var. *leightoniae* (G. G. Smith) Bayer in error).

H. cooperi var. *cooperi* clumps slowly from the base (more rapidly in some forms, like the former *H. vittata*, a light green, very bristled and pointed leaf variant), and makes handsome, flattish clumps of translucent tipped rosettes, with translucent bristles adorning the margins and keels. Forma *pilifera* is variable, but more inclined to sit deeper in the soil, showing just the rounded ends of the leaves in the wild, which are translucent and with less prominent bristles, but more inclined to colour purplish-red in full light, with the veining beneath the translucent end-area

showing up attractively. Var. *leightonii* has this character as already mentioned, and a habit midway between the aforementioned, sitting not so deeply as fa. *pilifera*, but not so vigorously growing or so inclined to open out as var. *cooperi*. The leaves are thicker and wider than the type and more abruptly pointed.

H. cooperi var. *cooperi* fa. *pilifera*

H. cooperi var. *cooperi* fa. *pilifera*

H. cooperi var. *cooperi* fa. *cooperi (vittata)*

H. cooperi var. *cooperi* fa. *pilifera*

H. cooperi var. *leightonii*

H. correcta

H. cordifolia Haw., *Suppl. Pl. Succ.* 60 (1819); *Haw. Handb.* 109 (1976), Bayer

There is little doubt that this species equates to *H. viscosa*, probably to the thicker leaved fa. *asperiuscula*.

H. correcta von Poelln., *Fedde's Repert. Spec. Nov.* 43:103 (1938); *Kakteenk.* 132 (1937)—as *H. blackburniae*; *Aloe* 11:8 (1973), C. L. Scott & 12:89 (1974), Bayer; *The Second Fifty Haw.* 28 (1975), Pilbeam; *Haw. Handb.* 109 (1976), Bayer; *Nat. Cact. Succ. Journ.* 34:28 (1979), Bayer

Bayer attributed this species to synonymy with *H. emelyae*. Its resemblance to forms of *H. turgida* shows the closeness of these species.

H. cuspidata Haw., *Suppl. Pl. Succ.* 51 (1819); *Cact. Amer.* 10:196 (1938), J. R. Brown; *The First Fifty Haw.* 15 (1970), Pilbeam; *Haw. Handb.* 110 (1976), Bayer

Although well-known in collections, and consistently of the same appearance, this name is rejected as being probably a hybrid between *H. cymbiformis* and *H. retusa* of garden origin. It has no real merit.

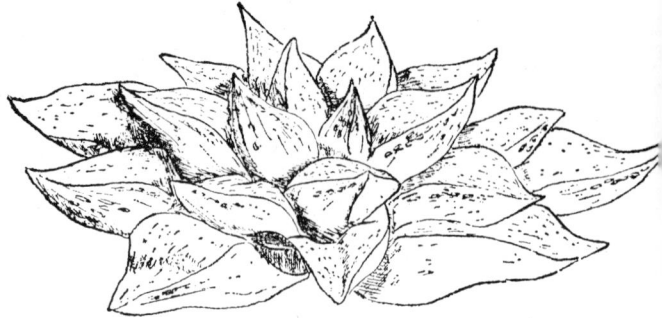

H. cuspidata

H. correcta

H. cuspidata

H. cymbiformis (Haw.) Duv., *Pl. Succ. Hort. Alenc.*
7 (1809); *Trans. Linn. Soc.* 7:8 (1804) as *Aloe
cymbiformis*, Haworth; *Syn. Pl. Succ.* 93 (1812),
Haworth; *Pflanz.* 4.38:101 non fig. 33A (1908) Berger;
Monogr. 11. 1 (1836–63), as *Aloe cymbaefolia*, Salm-
Dyck; *Cact. Amer.* 15:160 (1943), J. R. Brown; *The
First Fifty Haw.* 15 (1970), Pilbeam
Section Arachnoideae subsection Cymbifoliae

The plethora of varieties erected by Triebner and
von Poellnitz in the 1930s was scathingly attacked by
G. G. Smith as long ago as 1948, when he said 'I have
mentioned several instances of variable forms in
Haworthias, and I feel that if the erection of "new
species" and "new varieties" is to continue on such
slight differences, the position in regard to this genus is
going to be impossible. On krantzes around East
London and other parts of the country, one finds *H.
cymbiformis* or its varieties. On close examination, it
will be noted that in an area of a square yard, they take
on several different forms. There are the huge clusters
made up of many large rosettes with broad, thick,
entire leaves—the result of growing in a pocket of deep
rich soil, and getting possibly only the morning sun.
Then there are smaller clusters made up of very small
compact rosettes with proportionately smaller
leaves—growing in a pocket of shallow poor soil and
possibly in the full day's sun. Then there are the
loosely arranged clusters, made up of rosettes with
long, narrow, loosely arranged, toothed leaves—
growing in full shade, usually in rich soil. Must these
three forms be given distinctive names, simply because
they look different?'

H. cymbiformis var. *cymbiformis* fa. *cymbiformis*

In spite of this tirade, Jacobsen's listing in his
Handbook and Lexicon has ensured their continued
acceptance by those collectors seeking finer points of
variation in the genus, at least until recently when
Bayer's papers in various journals have made general
knowledge the variability of species of *Haworthia*.
Furthermore Bayer has concluded that there are other
'species' which have their allegiance with *H. cymbiformis*, with more valid distinction than the varieties

erected by Triebner and von Poellnitz, and later by
Resende. They are *H. planifolia*, *H. ramosa*, *H.
incurvula*, *H. umbraticola* and *H. planifolia* var.
transiens.

In order to maintain those of the former varieties
which have become widely accepted in collections as
distinctive and which maintain their individual form in
cultivation, a few are maintained here at form level, but
there is no doubt that no more than form status is
warranted for these variants in this complex species.
None of them present any difficulty in cultivation,
unless it is in containing them within bounds, as most
will quickly fill a wide pan with rosettes, so prolifically
do they offset from the base, on short stolons rooting as
they go. They are favourite candidates for Gordon
Rowley's 'Haworthia balls' treatment, i.e. tying
together two clumps back-to back to form a complete
ball of rosettes, which he then strings up like a hanging
basket.

Also equated here is von Poellnitz's *H. gracilidelineata*, which a contemporary drawing by J. T. Bates
(1933) shows clearly to be a highly translucent, small-
rosette, clustering species with allegiance here.

The species resolves as follows:

H. cymbiformis (Haw.) Duv. var. *cymbiformis*.

H. cymbiformis var. *cymbiformis* fa. *multifolia*
(Triebn.) Pilbeam, stat. nov. (syn. *H. cymbiformis* var.
multifolia Triebn., *Fedde's Repert. Spec. Nov.* 45:166
(1938));

H. cymbiformis var. *cymbiformis* fa. *obesa* (von
Poelln.) Pilbeam, stat. nov. (syn. *H. cymbiformis* var.
obesa von Poelln., *Fedde's Repert. Spec. Nov.* 45:165
(1938));

H. cymbiformis var. *cymbiformis* fa. *planifolia* (Haw.)
Pilbeam, comb. nov. (syn. *H. planifolia* Haw., *Phil.
Mag.* 44:282 (1825));

H. cymbiformis var. *cymbiformis* fa. *ramosa* (G. G.
Smith) Bayer (syn. *H. ramosa* G. G. Smith).

H. cymbiformis var. *cymbiformis* fa. *gracilidelineata*
(von Poelln.) Pilbeam, comb. nov. (syn. *H. gracilidelineata*, von Poelln., *Fedde's Repert. Spec. Nov.* 31:84
(1932); & 44:236 (1938), as *H. bilineata* var.
gracilidelineata).

H. cymbiformis var. *incurvula* (von Poelln.) Bayer
(syn. *H. incurvula* von Poelln.).

H. cymbiformis var. *transiens* (von Poelln.) Bayer
(syn. *H. planifolia* var. *transiens* von Poelln. and *H.
cymbiformis* var. *translucens* Triebn. & von Poelln.).

H. cymbiformis var. *umbraticola* (von Poelln.) Bayer
(syn. *H. umbraticola* von Poelln. and *H. hilliana* von
Poelln.).

These varieties and forms are distinguished as
follows:

The type, var. *cymbiformis* fa. *cymbiformis*, covers
the spectrum of former erections under this epithet,
apart from those singled out above. It invariably makes
large clumps of rosettes, of a light green, shining,
usually turgid appearance, turning yellow and orange-
red in full light. The leaves are 'cymbiform' (boat-

H. cymbiformis var. *cymbiformis*

H. cymbiformis var. *cymbiformis (brevifolia)*

H. cymbiformis var. *cymbiformis (compacta)*

H. cymbiformis var. *cymbiformis* fa. *gracilidelineata*

shaped) with usually a prominent keel in the upper part of the leaf, the lower surface convex and rounded, the upper surface flat to somewhat convex; there are more or less translucent flecks towards the tips of the leaves. The illustrations show the wide variation encountered in this species, and the following are the most distinctive.

H. cymbiformis var. *cymbiformis* fa. *cymbiformis* (variegated form)

fa. *multifolia* is retained as it is one of the most attractive for collectors, with many leaves to the rosette, and forming dense clumps of comparatively small rosettes, each only 3 or 4cm in diameter, with leaves more incurving than most others. It is taken by Bayer to include *H. lepida*, but plants in cultivation according with G. G. Smith's description are somewhat different-looking. Reported from Transkei, about 16 miles east Idutywa, Bathurst, although Bayer reports that it comes from the same locality as Triebner and von Poellnitz's var. *brevifolia*, at Hell's Gate, north of Uitenhage. There is some doubt as to the reported locality of *H. lepida*, along the Fish River, west of Fort Brown, and Bayer reports that it does not appear to exist any longer in the wild;

H. cymbiformis var. *cymbiformis* fa. *multifolia*

H. cymbiformis var. *cymbiformis* fa. *gracilidelineata*

H. cymbiformis var. *cymbiformis* fa. *multifolia*

H. cymbiformis var. *cymbiformis* fa. *obesa*

H. cymbiformis var. *cymbiformis* fa. *ramosa*

fa. *obesa* is retained as it is a distinct, very thick-leaved form. It was reported by von Poellnitz to come from the Transkei, about 16 miles east of Idutywa, Bathurst, the same reported locality as that for fa. *multifolia*, but Bayer reports it from the Bashee River in the Transkei. This form makes clumps of dense, closely knit rosettes, with the ends of the leaves giving the appearance of being inflated and squeezed out of the base of the rosette. The ends are rounded and have very little evidence of margins or keel, with prominent flecks giving them a dappled appearance.

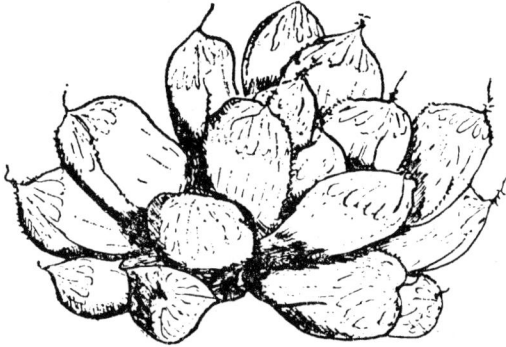

H. cymbiformis var. *cymbiformis* fa. *obesa*

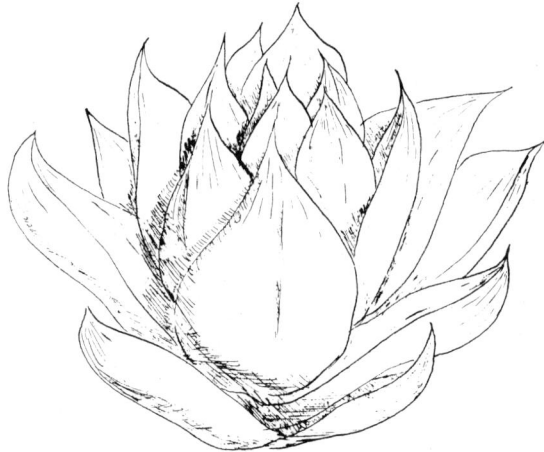

H. cymbiformis var. *cymbiformis* fa. *planifolia*

fa. *planifolia* is retained at form level to represent the broad, flat-leaved forms, which led Haworth to describe it as a separate species, and Triebner and von Poellnitz to rhapsodize on no less than 11 further varieties and forms. It says much for the lack of distinction between these that they are represented in collections usually by, at most, two or three. They are quite justifiably submerged under this form, which Bayer did not consider of sufficient distinction to maintain even at varietal level. It forms large rosettes

of 8 to 10cm diameter, more slowly forming massive clumps. The leaves are broad and flattened with little thickness, the 'cymbiform' shape becoming that of a flat-bottomed barge. Localities recorded for this form include Graaff Reinet (unlikely according to Bayer); Kowie River; Port Elizabeth; Hankey; Signal Hill near Grahamstown; Albany Division; Germiston in the Karoo; Bonnievale; near East London; Fort Beaumont; Sulphur Bath, 6 miles south-east of Fort Beaumont; Grahamstown; Cape Town, Quagga West; Baakens Valley near Port Elizabeth; and Prince Albert's Pass (this locality for var. *transiens*—see below).

H. cymbiformis var. *cymbiformis* fa. *planifolia*

fa. *ramosa* is distinguished from smaller forms of the type by its unique habit of forming elongated stems. Bayer reports that it is found only in an area north of Wooldridge near Peddie, 'at the east end of a vast south-facing (sunless) horseshoe of rock ... normal acaulescent forms (of *H. cymbiformis*) occur along the western end. It prefers less direct sunlight than most other forms.

H. cymbiformis var. *cymbiformis* fa. *ramosa*

fa. *gracilidelineata* is a well-known form in cultivation, usually not bearing a name at all, or sometimes seen labelled, wrongly, *H. translucens* or *H. pellucens*. It is an exceedingly pretty form, with almost wholly translucent leaves, incurving in the style of *H. cymbiformis* var. *incurvula*, with which it was contemporarily described, and with which it may well equate as a more translucent form (but it has page preference, and would be the preferred name if this were brought about). The rosettes are no more than 3cm or so wide, and it forms dense, closely-knit clumps.

var. *incurvula* occurs, according to Bayer, in a small area of fractured shale in Pluto's Vale east of Grahamstown, and is continuous, according to G. G. Smith (*Journ. S.A. Bot.* 16:1 (1950)) with the type in this area. Similar, if not identical plants occur at Andrieskraal, north of Humansdorp, which is a long way to the west from the Pluto's Vale locality, and 'there is an apparent, if obscure, relationship with *H. translucens* subsp. *tenera*, which bears further investigation', Bayer asserts. It is a narrow-leaved variety with incurving leaves, rounded and translucent at the tips.

H. cymbiformis var. *incurvula*

H. cymbiformis var. *incurvula*

var. *transiens*, a former variety of *H. planifolia*, and equating to the later described *H. cymbiformis* var. *translucens*, was reported from Prince Albert's Pass. Bayer maintains the name for the translucent-leaved plants found in this area. Exceedingly attractive clusters of rosettes are produced if sufficient light is given to keep the plants compact and the leaves close-clasping, but this is a variety which prefers less exposed conditions in habitat, and does not take too kindly to overmuch sun, so that a delicate balance must be struck to grow it at its best.

H. cymbiformis var. *transiens*

var. *umbraticola* is a variety recorded from the Swaartwater Poort west of Alicedale. The leaves are very obtuse, oblong and round in transverse cross section, with large clear areas at the ends of the leaves separated by lengthwise lines. If grown in good light so as to keep the plant compact this variety can make a wonderfully translucent cluster of leaf-tips, barely protruding above the soil in the wild, in the manner of some forms of *H. cooperi*.

H. cymbiformis var. *umbraticola*

H. cymbiformis var. *umbraticola*

H. decipiens von Poelln., *Fedde's Repert. Spec. Nov.* 28:103 (1930); *Cact. Amer.* 37:127 (1965), J. R. Brown; *The Second Fifty Haw.* 29 (1975), Pilbeam; *Haw. Handb.* 111 (1976), Bayer
Section Arachnoideae subsection Arachnoideae

   This species makes a large flat rosette (about 7cm across or more), with leaves smooth and shining, lying flat from the centre and incurving at the tips like a water-lily flower in full sunshine. It is a distinctive species with strong, white, cartilaginous teeth on the margins (not on the keel), widely spaced and 2 to 2.5mm long. Its colouring in the leaf varies from dark grey-green when mature to a shining, light green in younger plants, with darker lines beneath the surface length-wise and tessellately connecting. It colours red and yellow in full light if allowed to go a little thirsty. It was reported from the Prince Albert District near the Zwartberg mountains. It grows lustily from seed (hardly believing its luck I suspect, as the odd habitat plants I have seen have obviously had a hard time, and have hardly any green coloration, looking a roasted red-orange, with dried outer leaves and leaf-tips). It does not respond kindly to overwatering at any time once it has made its thick, below-soil-level stem, and water in the centre of the rosette will ensure its rapid demise in greenhouse conditions. Seedlings, however, seem to appreciate the moisture and put on weight, like overfed children, almost visibly. Enough sunlight to bring out some of the autumn-like colouring suits them best, with watering around rather than over the top of the rosette.

H. decipiens

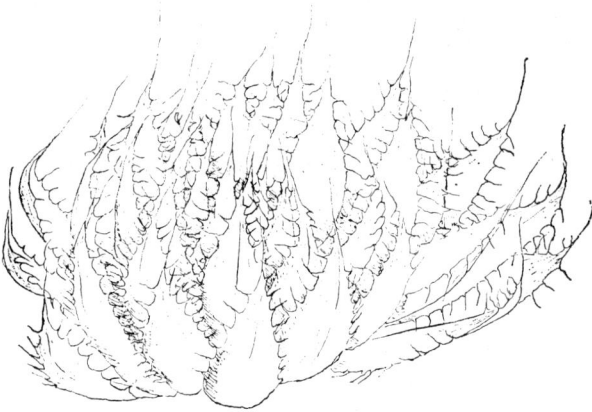

H. decipiens

H. dekenahii G. G. Smith, *Journ. S.A. Bot.* 10:140 (1944); *Haw. Handb.* 111 (1976), Bayer; *New Haw. Handb.* 53 (1982), Bayer

Bayer regards this species as a form of *H. retusa*, 'inextricably involved with *H. turgida*'. It is known only from a very restricted locality near Albertinia. While the type in his 1976 book was discarded, with the inference to be drawn that it was believed to be a

natural hybrid between the two above-mentioned species, the variety named by G. G. Smith, var. *argenteo-maculosa*, was referred to form status under *H. retusa* since 'this is clearly an eastern form of *H. retusa* with many silvery flecks in the leaf end-area'. In his 1982 book Bayer turned around and reinstated the name *H. dekenahii* as a variety of *H. retusa*, reducing '*argenteo-maculosa*' to synonymy beneath it.

H. divergens Bayer, *Haw. Handb.* 113 (1976)
Section Loratae subsection Loratae

This species was described for the first time in Bayer's *Haworthia Handbook*, but has not found its way into collections as yet. A small plant received at the time it was described, or just before, has made a slow, incurving rosette of toothed, narrow leaves. It is named for its divergence from two closely related species: *H. angustifolia*, from which it is distinct by its conspicuous teeth and a geographical break of about 160km; and *H. variegata*, from which it is separated by the Langeberg mountain range, the Gouritz river valley and coastal vegetation including *H. chloracantha* in the Mossel Bay area. It differs from both by its incurved leaf-tips. Bayer suggests that it may also intergrade with *H. zantnerana*. It has attenuated leaves up to 5cm long, and narrow (about 6mm), without

H. divergens

much substance (3mm thick). In the upper third they are a little variegated. Its relationship suggests careful watering and moderately good light conditions, although over exposure will redden the plant and may cause shrivelling.

Reported from an area north of the Outeniqua mountains between Oudtshoorn and Willowmore.

H. eilyae von Poelln., *Kakteenk.* 152 (1937); *Fedde's Repert. Spec. Nov.* 43:108 (1938); *Cact. Amer.* 10:124 (1939), J. R. Brown; *Haw. Handb.* 114 (1976), Bayer

Referred to synonymy with *H. glauca* by Bayer, this species erected by von Poellnitz, as well as the variety described by Resende, var. *zantnerana*, is just a smaller-growing, more tuberculate form of var. *herrei*, not warranting recognition.

H. emelyae von Poelln., *Fedde's Repert. Spec. Nov.* 42:271 (1937); *Des. Pl. Life* 10:126 (1938) von Poellnitz—as *H. picta*; *Cact. Amer.* 11:174 (1940), J. R. Brown—as *H. picta*; *The Second Fifty Haw.* 44 (1975), Pilbeam—as *H. picta*; *Excelsa* 5:86 (1975) Bayer; *Nat. Cact. Succ. Journ.* 34:28 (1979), Bayer Section Retusae subsection Retusae

The name *H. picta* seems now to have finally been accepted by enthusiasts for this genus as synonymous with this species, and so, by priority, the name *H. emelyae* is preferred. As the alternative name implies it is a colourful species, 'picta' meaning painted, the end-area of the leaves becoming a palate of bronze, lilac, pink and red in full sunlight, set off wonderfully by the low, translucent tubercles that cover this area. The leaves fit tightly together to form a low, flat-topped rosette, clustering reluctantly from the base in cultivation in time. It is not quick-growing, although it will reach maturity in about four years from seed, and seedlings grow readily enough, if slowly. Bayer regards *H. correcta* as synonymous with this species, and the closeness of *H. correcta* to *H. turgida* in appearance, if not geography, emphasizes possible relationships between all three. In the last-mentioned article by Bayer, accompanying photographs show the considerable variation of the species, and a new variety is erected: var. *multifolia*, differing from the type by the completely different-looking little recurving leaves. Bayer equates it to the species *H. emelyae* by virtue of its occurrence in the wild, and its flowers. As the name indicates it has many more leaves than most forms seen of this species, which are narrow and sharply pointed, up to 45mm long, 10mm wide, 8mm thick at the middle, shaped very much like the leaves of *H. serrata* or *H. heidelbergensis*.

H. emelyae var. *emelyae*

H. emelyae var. *emelyae* (from Uniondale)   *H. emelyae* var. *emelyae*

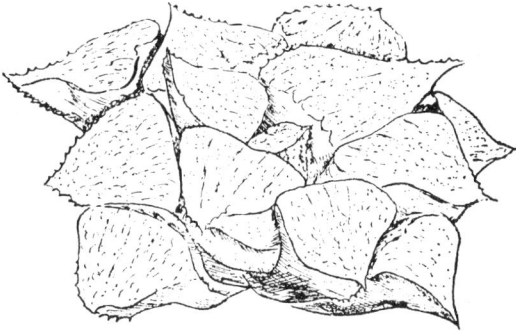

H. emelyae var. *emelyae*

This species occurs north of the Langeberg mountain range, and is reported by Bayer: from between Calitzdorp and Vanwyksdorp; north-east of Garcias Pass; near the locality for *H. magnifica* var. *major* at Muiskraal; 20 miles west of Ladismith (this locality far from others for this species); in the Oudtshoorn area; in the vicinity of the Moeras River; from between Oudtshoorn and Mossel Bay; from Dysseldorp, east of Oudtshoorn; from Uniondale. The variety *multifolia* comes from Springfontein in the Riversdale District, a few miles west of Muiskraal.

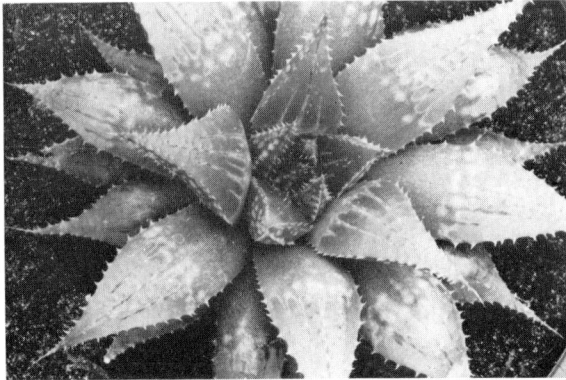

H. emelyae var. *multifolia*

H. fasciata (Willd.) Haw., *Berlin Mag.* 5:270 (1811), Willdenow; *Revis. Pl. Succ.* 54 (1821), Haworth; *Monogr.* 6:15 (1836–63), Salm-Dyck; *Cact. Journ.* 7:74 (1938), von Poellnitz; *Cact. Amer.* 10:19 (1938), J. R. Brown; *Succulenta* 34:38 (1952), Uitewaal; *The First Fifty Haw.* 18 (1970), Pilbeam; *Haw. Handb.* 116 (1976), Bayer
Section Coarctatae subsection Coarctatae

This is *not* the species so often seen under this label (lamentably too in many popular books) with long, flexible, attenuated leaves inclined to recurve and clustering strongly from the base, which is in fact *H. attenuata* fa. *caespitosa*. *H. fasciata* is a more slow-growing species with more affinity to *H. coarctata* in that the leaves are rather short and long-triangular, constantly incurving, 3 to 4cm long and about 1cm broad on an eventually tall-stemmed plant. It has white tubercles in more or less continuous bands across the backs of the leaves, but the upper surface is almost smooth and devoid of tubercles; the leaves are more rigid than *H. attenuata*, and usually a lighter, greyish-green, with less shine.

H. fasciata fa. *browniana*

H. fasciata fa. *fasciata*

The older named forms of this species are lost in obscurity, and appear in the main to have been based on differences observed on plants in cultivation whose origin was not recorded. They are, at most, small differences (size in fa. *major*, and colouring in fa. *perviridis*) and have no importance. Bayer reduces the species *H. browniana* to form status under this species, after investigation in the field. This form, found in only one locality, 6 miles south of Uitenhage, seems to have originated from only one clone (this revealed by cytological study by Dr. Peter Brandham of the Jodrell Laboratory, Royal Botanic Gardens, Kew—see *Kew Bull.* 28:348 (1973)). It is not rapid in growth, but will

H. fasciata fa. *fasciata*

eventually make a far larger plant than von Poellnitz envisaged in his original description (where he recorded the stem as 2cm long) making a thick column to about 12cm tall, clustering from the base and lower stem. The leaves are thick, and have tubercles more or less coalescing into thin, white lines across the backs of the leaves, with a base colour similar to the type, of greyish-green.

The only other form maintained here, which Bayer did not recognize, is the easily differentiated fa. *sparsa*, from the Witteklip mountains. This is short-leaved, and has quite separate tubercles on the backs of the leaves. It eventually makes a columnar plant, but is slow to do so, and does not make the height of the type or of fa. *browniana*.

The species resolves as follows:

H. fasciata (Willd.) Haw., fa. *fasciata*, the type, reported by Bayer from around Port Elizabeth and the Gamtoos Valley;

H. fasciata fa. *browniana* (von Poelln.) Bayer;

H. fasciata fa. *sparsa* von Poelln.

Other varieties and forms not recognized are: var. *concolor* Salm-Dyck, with concolorous tubercles; var. *perviridis*, entirely green; var. or fa. *major* (Salm-Dyck) von Poelln., a larger growing form; fa. *ovato-lanceolata* von Poelln., with broader leaves, maybe equating to fa. *browniana*; fa. *subconfluens* von Poelln. with narrow leaves; fa. *vanstaadensis*, little different from the type; fa. *variabilis* von Poelln., variably marked leaves, some with almost no tubercles.

H. ferox von Poelln., *Fedde's Repert. Spec. Nov.* 31:84 (1933); *Haw. Handb.* 116 (1976), Bayer.

There seems little doubt that Bayer is correct in maintaining that the plant von Poellnitz originally described was *Aloe humilis*, a somewhat Haworthia-like species of *Aloe* in superficial appearance. There have been several candidates offered by nurserymen over the years as this species, which have turned out to be *H. herbacea*, or *H. magnifica* var. *paradoxa*, or, according to Bayer, *H. turgida* in more robust forms, or *H. aristata* (in its more denticulate form).

H. fasciata fa. *sparsa*

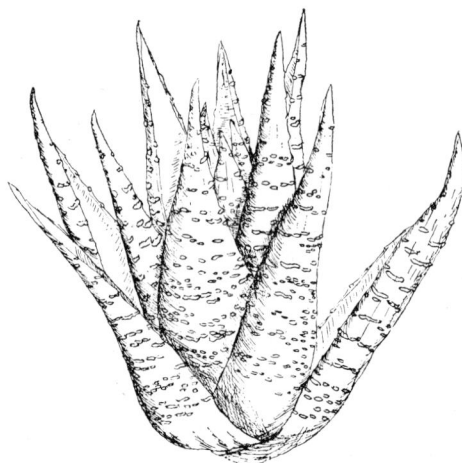

H. fasciata fa. *browniana*

H. cymbiformis var. *cymbiformis* (painting by Mel Roberts)

H. limifolia var. *striata* (painting by Mel Roberts)

H. venosa subsp. *tessellata* (painting by Mel Roberts)

H. archeri var. *archeri*

H. attenuata fa. *caespitosa*

H. bolusii var. *blackbeardiana*

H. bruynsii

H. coarctata (*greenii*)

H. coarctata subsp. *coarctata* var. *tenuis*

H. coarctata subsp. *adelaidensis*

H. comptoniana fa. *major* from Georgida

H. cooperi var. *cooperi* fa. *pilifera*

H. cymbiformis var. *incurvula*

H. cymbiformis (variegate)

H. decipiens

H. divergens

H. emelyae var. *emelyae* from Van Wyksdorp

H. floribunda from Riversdale

H. habdomadis var. *habdomadis*

H. herbacea

H. koelmaniorum

H. limifolia var. *gigantea*

H. lockwoodii from Laingsburg

H. maculata from Worcester

H. magnifica var. *magnifica* from Heidelberg

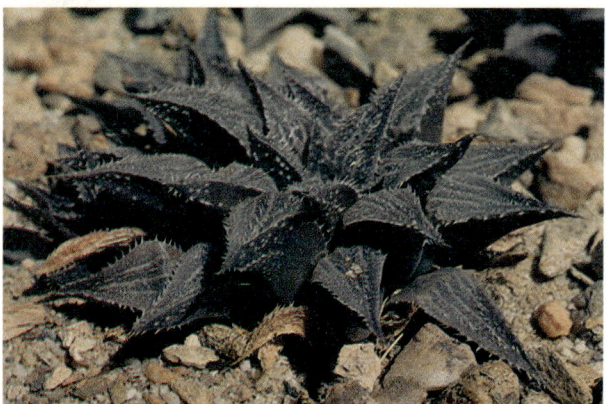
H. magnifica var. *major* from Riversdale

H. magnifica var. *maraisii* from Robertson

H. magnifica var. *meiringii* from Bonnievale

H. marginata

H. minima

H. mirabilis subsp. *mirabilis* fa. *beukmannii*

H. mirabilis subsp. *mirabilis* fa. *napierensis*

H. mirabilis subsp. *mirabilis* fa. *rubrodentata*

H. mirabilis subsp. *badia*

H. mirabilis subsp. *mundula* from Muiskraal

H. mutica from Bredasdorp

H. nigra fa. *angustata* from Thomas River

H. nortieri var. *nortieri*

H. nortieri var. *globosiflora* from Botterkloof

H. parksiana

H. poellnitziana

H. pumila

H. pygmaea fa. *major* from Great Brak

H. reinwardtii var. *brevicula*

H. reinwardtii var. *reinwardtii* fa. *chalumnensis*

H. reinwardtii var. *reinwardtii* fa. *kaffirdriftensis*

H. reinwardtii var. *reinwardtii* fa. *zebrina*

H. reticulata var. *hurlingii* near Bonnievale

H. retusa var. *retusa* fa. *retusa* from Albertinia

H. retusa var. *acuminata* from Gouritz

H. retusa var. *dekenahii*

H. rycroftiana

H. semiviva from Frazerburg

H. serrata

H. subattenuata (variegate)

H. truncata fa. *truncata*

H. ubomboensis

H. venosa subsp. *venosa* by the Breede River

Astroloba foliolosa

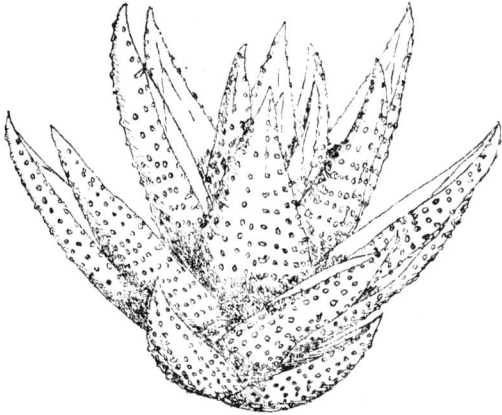

H. fasciata fa. *sparsa*

H. floribunda

H. floribunda von Poelln., *Fedde's Repert. Spec. Nov.* 40:149 (1936); *Cact. Amer.* 19:171 (1947), J. R. Brown; *The Second Fifty Haw.* 33 (1975) Pilbeam; *Haw. Handb.* 117 (1976), Bayer
Secton Loratae subsection Loratae

This is not a species often seen in collections, and does not lend itself to easy cultivation at all. It varies considerably in leaf-shape, but the common characteristic is the blunt, rounded end to the thin, dull, strap-like leaves, which curve this way and that, smooth-edged and minutely—or sometimes distinctly—toothed. It was named for its free production of flowering stems, four or five normally appearing more or less simultaneously, thin and wiry with few flowers. The more toothed forms show in addition a tendency to spotting on the lower parts of the leaves. It does in time produce basal offsets in cultivation.

Reported from north of Heidelberg on river banks in partial shade of 'Rhenoster' bushes and with only the

H. floribunda

tips of the leaves above the soil; and from around Riversdale. Bayer is of the opinion that this species is probably related to *H. parksiana*. Both grow in shaded conditions among mosses and lichens. A natural hybrid is reported where this species meets *H. retusa* fa. *longebracteata*.

H. floribunda

H. floribunda × *H. retusa* fa. *longebracteata* (natural hybrid)

H. fouchei von. Poelln., *Succulenta* 22:28 (1940); *Cact. Amer.* 41:59 (1969), J. R. Brown; *Haw. Handb.* 117 (1976) Bayer

Bayer is of the opinion that this species merges with *H. geraldii*, and in turn with *H. retusa*. Since it is widespread in cultivation and has a quite beautiful translucence and distinctive form it is maintained here at form level—see under *H. retusa*.

H. fulva G. G. Smith, *Journ. S.A. Bot.* 9:101 (1943); *Cact. Amer.* 25:31 (1953), J. R. Brown; *Nat. Cact. Succ. Journ.* 28:80 (1973), Bayer; *The First Fifty Haw.*, 18 (1970) Pilbeam; *Haw. Handb.* 117 (1976), Bayer

Referred by Bayer to synonymy with *H. coarctata*.

H. geraldii C. L. Scott, *Journ. S.A. Bot.* 31:123 (1965); *Haw. Handb.* 117 (1976), Bayer; *Excelsa* 5:84 (1975), Bayer

Bayer has indicated that this species should properly be regarded as synonymous with *H. retusa*, being merely a lighter green, vigorously clumping form of that species. It is maintained at form level.

H. glabrata (Salm-Dyck) Bak., *Trans. Linn. Soc.* 18:206 (1880); *Hort. Dyck.* 325 (1834), Salm-Dyck Section Coarctatae subsection Attenuatae

This is the often seen, very ordinary looking, green species with long, recurving, attenuated leaves, doing little to win collectors to this genus. The lower surface of the leaves is rough, from many small, concolorous tubercles. It presents no difficulty in cultivation, making a tangled clump of rosettes fairly quickly whatever its treatment. It is best grown in full light to darken or redden the somewhat pallid leaves and keep them short, when it looks its best.

Bayer had been unable to locate it in the field and has expressed the opinion that it may be a garden hybrid, but Dr Peter Brandham of Jodrell Laboratory, Royal Botanic Gardens, Kew, disputes this on the basis of cytological examination.

H. glabrata

H. glauca Bak., *Journ. Linn. Soc.* 18:203 (1880); *Cact. Amer.* 13:35 (1941), J. R. Brown; *The First Fifty Haw.* 18 (1970), Pilbeam; *Haw. Handb.* 118 (1976), Bayer

As originally described this is the stiffest-leaved of the group of species which Bayer has now sunk into synonymy with it (*H. herrei*, *H. jacobseniana*, *H. jonesiae*, *H. eilyae* and *H. armstrongii*). Bayer reduces two of these former species to variety and form level, but has acknowledged (in letters) the distinctive forms other than those retained, the more outstanding of which are maintained hereunder at form level. *H. armstrongii* is maintained at species level. The species resolves as follows:

H. glauca Bak. var. *glauca*.

H. glauca var. *herrei* (von Poelln.) Bayer.

H. glauca var. *herrei* fa. *jacobseniana* (von Poelln.) Pilbeam, stat. nov. (syn. *H. jacobseniana* von Poelln. in *Des. Pl. Life* 9:102 (1937)).

H. glauca var. *herrei* fa. *jonesiae* (von Poelln.)

Pilbeam, stat. nov. (syn. *H. jonesiae* von Poelln. in *Kakteenk.* 9:153 (1937)).

The type, *H. glauca* var. *glauca*, forms clumps of spiky, leaf-clothed upright stems to about 10 or 12cm tall, and 4 to 5cm wide, clustering readily from the base and below soil-level. The leaves are greyish-green, turning to dusky red in strong light or dry conditions, with no prominent tubercles, but with several ridges running lengthwise on the backs of the leaves, more pronounced when the plants are dry. *H. glauca* var. *herrei* embraces the less substantial, more flexibly leaved forms of the species, including those reduced to synonymy, which formerly had specific status; these according to Bayer are more westerly than the type, and somewhat more tuberculate as well as having more erect to erect-spreading leaves. As originally described var. *herrei* is quite large-growing, with stems to 20cm or longer, and 4cm or more wide, sprawling after achieving about 10cm in height, and clustering readily from the base; var. *herrei* fa. *jacobseniana* is the smallest of the forms, with stems no more than 8 to 10cm tall and strongly clustering to form low clumps of thickly-grouped stems, blue-green and most attractive, with irregular rows of tubercles on the backs of the short leaves (this form occurs at Miller near Jansenville); var. *herrei* fa. *jonesiae* is again very individual and attractive for collectors, with smooth, tubercle-less

short leaves, with slender stems, about 2 to 3cm wide, and rarely more than 12 or 15cm tall; the colouring is a subtle pale blue-green (this form was collected at Steytlerville). Former species *H. eilyae* equates more nearly to *H. glauca* var. *herrei*, and is not maintained separately.

H. glauca var. *glauca*

H. glauca var. *glauca*

H. glauca var. *herrei* fa. *herrei*

H. glauca var. *herrei* fa. *jacobseniana*

H. glauca var. *herrei* fa. *jonesiae*

H. globosiflora G. G. Smith, *Journ. S.A. Bot.* 16:11 (1950); *Haw. Handb.* 119 (1976), Bayer

Bayer reduces this species to varietal level beneath *H. nortieri*.

H. gracilidelineata von Poelln., *Fedde's Repert. Spec. Nov.* 31:84 (1932); *Haw. Handb.* 119 (1976), Bayer

A contemporary drawing by J. T. Bates clearly shows this name to be applied to a plant in cultivation with small rosettes and very translucent leaves with clear alliance to *H. cymbiformis*. Because of its distinctive form it is upheld at form level—see *H. cymbiformis*.

H. gracilis von Poelln., *Fedde's Repert. Spec. Nov.* 27:133 (1929); *Des. Pl. Life* 9:90 (1937); *Cact. Amer.* 27:154 (1955), J. R. Brown; *Haw. Handb.* 119 (1976), Bayer

Bayer refers this species to synonymy with *H. translucens*.

H. graminifolia G. G. Smith, *Journ. S.A. Bot.* 8:247 (1942); *Haw. Handb.* 120 (1976), Bayer
Section Loratae subsection Fusiformes

With *H. blackburniae* this species shares the distinction of having tuberous roots, for which the section Fusiformes was erected by Barker. The foliage in this species is even narrower and more like thin blades of grass than *H. blackburniae*, and is usually much longer, to 30cm and only about 1.5mm wide, channelled, rough to the touch from many marginal teeth, drying readily at the tips. It was found near the Cango Caves, north of Oudtshoorn, among tall, coarse grass on south-west facing slopes, and was difficult to find even when in flower, so grass-like is its appearance. Good drainage is needed to ensure the plants do not rot, which they are inclined to do in cultivation. Seeds have been produced from collected plants in cultivation, which have resulted in some seedlings being made available to enthusiasts for Haworthias, surely the only people interested in possessing this 'grass'—but then many plants of *Calibanus hookeri* are grown in cultivation, which I regard as far less attractive than *H. graminifolia*.

H. graminifolia

H. granulata Marl., *Trans. Roy. Soc. S. Afr.* 2:39 (1910); *Haw. Handb.* 120 (1976), Bayer
Referred by Bayer to *H. venosa*.

H. greenii Bak., *Journ. Linn. Soc.* 18:202 (1880); *Haw. Handb.* 120 (1976), Bayer
With its varieties and forms this species is reduced to synonymy with *H. coarctata*.

H. guttata Uitew., *Des. Pl. Life* 19:136 (1947); *Haw. Handb.* 121 (1976), Bayer

Bayer equates this species to *H. reticulata* or possibly a hybrid between *H. reticulata* and *H. magnifica* var. *maraisii*. Plants often seen in collections under this name are the former *H. schuldtiana* var. *erecta*, reduced by Bayer to synonymy with *H. magnifica* but these bear little resemblance to what was originally described as *H. guttata*, a much daintier plant.

H. haageana von Poelln., *Fedde's Repert. Spec. Nov.* 28:104 (1930); *Cact. Amer.* 12:107 (1940), J. R. Brown; *The First Fifty Haw.* 18 (1970), Pilbeam; *Cact. Succ. Journ. NSW.* 8:9 (1971), Bayer; *Nat. Cact. Succ. Journ.* 27:10 (1972), Bayer; *Haw. Handb.* 121 (1976), Bayer
Referred with the variety *subreticulata* to *H. reticulata*.

H. habdomadis von Poelln., *Fedde's Repert. Spec. Nov.* 46:271 (1939); *Des. Pl. Life* 11:88 (1939); *The Second Fifty Haw.* 34 (1975), Pilbeam; *Haw. Handb.* 121 (1976), & see correction in *Nat. Cact. Succ. Journ.* 32:18 (1977), Bayer
Section Arachnoideae subsection Limpidae
Bayer in his Handbook reduced this species to varietal status beneath *H. inconfluens*, but corrected this subsequently since it contravened the International Code for Botanical Nomenclature article 60, which states that a name does not have priority outside its own rank, so that '*inconfluens*' at form level could not take priority over *H. habdomadis*, erected by von Poellnitz as a species. Bayer rejected the specific name *H. mucronata*, described by Haworth in 1819, as indeterminate, which might have had preference here, and resolved the species as follows:
   H. habdomadis von Poelln. var. *habdomadis*.
   H. habdomadis var. *inconfluens* (von Poelln.) Bayer, (syn. *H. altilinea* var. *limpida* fa. *inconfluens* von Poelln. & *H. mucronata* var. *mucronata* fa. *inconfluens* von Poelln.).
   H. habdomadis var. *morrisiae* (von Poelln.) Bayer, (syn. *H. altilinea* var. *morrisiae* von Poelln. & *H. mucronata* var. *morrisiae* von Poelln.).

H. habdomadis var. *habdomadis*

The type, var. *habdomadis*, was described from plants collected at Seweweekspoort, and Bayer reports it from here and from other areas in quartzitic soils, particularly from Dwarsriver, west of Ladismith. From the collector's point of view it is the most attractive of the varieties, with its sometimes quite heavy armour of translucent teeth on the margins and keel (sometimes there are two bristled keels present on the wide leaves). Apart from a tendency to rot if water is allowed to lie in the dense rosette of leaves, and an agonizingly slow rate of growth, it presents no problems in cultivation, although its growing period is short, no more than a few months in the year. When not growing it closes each rosette into a bristly ball, and reduces its diameter to half its turgid size.

H. habdomadis var. *inconfluens* occurs in distinct populations around Ladismith. Its nearest relation, apart from its sibling varieties, seems to be *H. lockwoodii*, to which it bears a superficial similarity but with narrower leaves. It is a beautiful variety with its incurving, sharply pointed leaves, the tips gathered at the centre, with occasionally very small teeth on the margins of the leaves, but usually untoothed, the rosette about 5 cm wide.

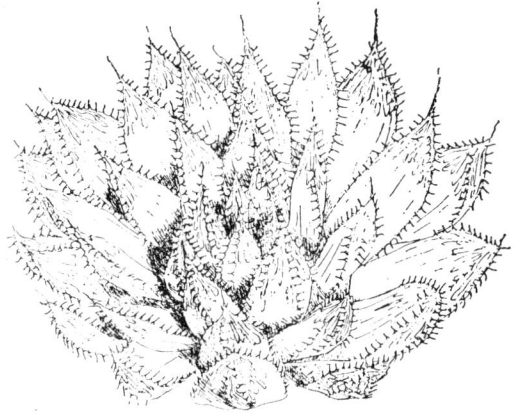

H. habdomadis var. *habdomadis*

H. habdomadis var. *inconfluens*

H. *habdomadis* var. *morrisiae* is a clearly identifiable variation found in the Calitzdorp to Oudtshoorn area, deeply seated in the ground and usually solitary. It is characterized by a yellowish-green, almost chlorotic appearance, with narrower and shorter, tightly packed, incurving leaves, usually without teeth or with very few, very small ones at the margins on young leaves.

H. heidelbergensis G. G. Smith, *Journ. S.A. Bot.* 14:42 (1948); *Cact. Amer.* 25:60 (1963), J. R. Brown; *The Second Fifty Haw.* 36 (1975), Pilbeam; *Haw. Handb.* 121 (1976), Bayer; *Aloe* 11:8 (1973), C. L. Scott & 12:89 (1974), Bayer
Section Retusae subsection Retusae
    This is a charming, small species in this section of attractive species. The rosette is stemless and at maximum size about 7cm wide, suckering from around the base to form small, slow clumps. It is surprising in such a small rosette to find that there are about 35 leaves when full grown, the younger leaves green and erect, becoming darker as they age and grow more outwards, recurving with a tendency to incurve at the tips. The leaves are up to 35mm long, 8mm broad at

H. *habdomadis* var. *morrisiae*

the widest part, 6mm at the base. The end-area is semi-transparent with only 3 to 4 longitudinal lines, the centre one longer and often extending to the tip of the leaf, where there is a whitish, transparent, terminal bristle. The margins bear small teeth, more prominent in the middle part, and there are small teeth on the keel, with sometimes two keels present. It presents no problems in cultivation, but is slower to form clumps than many in this section. It looks better for being grown in fairly strong light, when there are contrasting dark and light greens apparent. It comes from a locality immediately east of Heidelberg, and according to Bayer 'bears no obvious direct affinities with any other species in the retuse complex'.

H. helmiae von Poelln., *Fedde's Repert. Spec. Nov.* 41:201 (1937) & 44:223 (1938); *Cact. Journ.* 6:18 (1937); *Cact. Amer.* 18:38 (1946), J. R. Brown (photographs incorrect); *Haw. Handb.* 121 (1976), Bayer
    Referred to *H. aristata* at varietal level.

H. heidelbergensis

H. heidelbergensis

H. henriquesii Res., Mem. Soc. Brot. 2:150 (1941)
Dubiously identifiable, this species seems to fall under *H. coarctata*.

H. herbacea (Mill.) Stearn, *Cact. Journ.* 7:40 (1938); Miller, *Gard. Dict.* ed. 8 no. 18 (1768)—as *Aloe herbacea*; *Monogr.* 10:2 (1836–63), Salm-Dyck; *Second Fifty Haw.* 5–8 (1975), Pilbeam; *Nat. Cact. Succ. Journ.* 27:51 (1972), Bayer; *Haw. Handb.* 122 (1976), Bayer; non Scott, *Cact. Amer.* 49:205 (1977)
Section Retusae subsection Turgidae
This species has been the subject of intensive study in the field by Bayer. His paper in the National Cactus & Succulent Society's journal (1972 ref. above) on this species is excellent reading. He reduces to synonymy several species which were set up in the 1930s often on the basis of a single plant and without proper study in the field. Species reduced to synonymy are *H. aegrota*, *H. pallida* and var. *paynei*, *H. submaculata* and *H. luteorosea*. There is a good illustration of *H. herbacea* in Salm-Dyck's monograph on Aloes and Mesembryanthemums (as *Aloe atrovirens* DC).

As may be guessed from the species reduced to synonymy this is a widespread and variable species, but apart from size not so variable as might be supposed: the rosettes may be anything from 2 or 3cm across (*H. pallida* var. *paynei*) to 6 or 7cm; their colouring varies from a pale, chlorotic green (*H. aegrota*) through mid-green to quite dark green; the flower-colour varies from white with grey-brown stripes on the outside of the petals to salmon-pink with almost maroon stripes (*H. luteorosea*). But the overall shape and habit of the species is constant, with erect to incurving leaves, the upper surface inflated and convex, suddenly tapering, lower surface rounded, keeled in the upper part, with many translucent teeth on the margins and keel and sometimes on the surface of the leaves; both surfaces are peppered with translucent flecks, more so towards the tip of the leaf, often surmounted by translucent teeth. Small clumps of rosettes are formed, but it takes some years to make a clump the size of a man's fist, and flowers are produced in England in the spring. Plants will turn yellow if exposed to a great deal of sunlight, although a good amount of light is necessary to maintain a close rosette, and to develop the bristles and translucent flecks to the full. Cultivation is not difficult, but growth is slower than the look of the plant would suggest.

H. herbacea

H. herbacea

Scott's application of the name *H. arachnoidea* to plants generally accepted as this species is believed to be wrong. Examination of contemporary illustrations at the Herbarium Library, Kew, and elsewhere, support Bayer's retention of the name *H. herbacea* and its application.

Reported from Ribbokkop in the Cape area.

H. herrei von Poelln., *Fedde's Repert. Spec. Nov.* 26:24 (1929); *Cact. Amer.* 13:97 (1941), J. R. Brown; *Haw. Handb.* 122 (1976), Bayer

Bayer has placed this species under *H. glauca* as a variety.

H. hurlingii von Poelln., *Fedde's Repert. Spec. Nov.* 41:202 (1937); *Nat. Cact. Succ. Journ.* 27:10 (1972), Bayer; *Second Fifty Haw.* 5–8 (1975), Pilbeam; *Haw. Handb.* 123 (1976), Bayer

Referred as a variety to *H. reticulata*.

H. hybrida (Salm-Dyck) Haw., *Revis. Pl. Succ.* 51 (1821); *Cat. Rais.* 41 (1817), Salm-Dyck

A self-confessed hybrid, hardly warranting the speculation hitherto as to its parentage, which was the result no doubt of some Georgian conservatory sinning.

H. icosiphylla Bak., *Journ. Linn. Soc.* 18:207 (1880)

Another garden hybrid, which should long ago have disappeared from listings of the genus. It is of unprepossessing appearance, owing allegiance probably to *H. glabrata* or *H. tortuosa* in its parentage, both to some extent dubious species themselves.

H. inconfluens (von Poelln.) Bayer, *Haw. Handb.* 123 (1976), Bayer; *Fedde's Repert. Spec. Nov.* 45:169 (1938)—as *H. altilinea* var. *limpida* fa. *inconfluens*; and 49:29 (1940)—as *H. mucronata* var. *limpida* fa. *inconfluens*, von Poellnitz

The brief sunrise of this as a species was quickly dimmed by an erratum slip to Bayer's *Haworthia Handbook*, reducing it to varietal level beneath the prior specific name *H. habdomadis*—see under the latter name.

H. incurvula von Poelln., *Fedde's Repert. Spec. Nov.* 31:85 (1932); *Des. Pl. Life* 8:45 (1938); *Cact. Amer.* 23:42 (1951), J. R. Brown; *The First Fifty Haw.* 20 (1970), Pilbeam; *Journ. S.A. Bot.* 16:1 (1950), G. G. Smith

Bayer has placed this species under *H. cymbiformis*.

H. integra von Poelln., *Fedde's Repert. Spec. Nov.* 33:239 (1933); *Haw. Handb.* 125 (1976), Bayer

With no real information as to its habitat (given simply as Little Karoo) the true identity of this species is in doubt. Von Poellnitz likened it to *H. reticulata* and *H. incurvula*. Plants in circulation under this name are more slender and more lanceolate-leaved than these

species, with reddish, longitudinal lines beneath the surface, but they are not verifiable.

H. intermedia von Poelln., *Kakteenk.* 134 (1937); *Haw. Handb.* 125 (1976), Bayer

Bayer relegates this species to synonymy with *H. reticulata*.

H. isabellae von Poelln., *Fedde's Repert. Spec. Nov.* 44:226 (1938); *Haw. Handb.* 125 (1976), Bayer

This is an obscure species based on a description of a single plant received by von Poellnitz in 1938 from near Port Elizabeth. Bayer dismisses it as a variant of *H. translucens*, on the basis of the locality data.

H. jacobseniana von Poelln., *Des. Pl. Life* 9:102 (1937); *Fedde's Repert. Spec. Nov.* 43:109 (1938), von Poelln.; *Cact. Amer.* 23:72 (1951), J. R. Brown; *Haw. Handb.* 125 (1976), Bayer

This species falls within the concept of *H. glauca* according to Bayer. It is here maintained at form level—see under that species.

H. janseana Uitew., *Cact. Vetpl.* 6:45 (1940); *Cact. Amer.* 34:13 (1962), J. R. Brown; *Haw. Handb.* 125 (1976), Bayer

A plant of unknown origin, thought probably to be a hybrid near *H. turgida*. Plants in cultivation under this name are a form of *H. angustifolia*.

H. jonesiae von Poelln., *Kakteenk.* 9:153 (1937); *Fedde's Repert. Spec. Nov.* 43:109 (1938); *Cact. Amer* 22:60 (1950), J. R. Brown; *Haw. Handb.* 125 (1976), Bayer

Bayer has reduced this species to synonymy with *H. glauca*. It is retained under that species here at form level.

H. cv. kewensis von Poelln., *Fedde's Repert. Spec. Nov.* 49:57 (1940); *Cact. Amer.* 28:191 (1956), J. R. Brown; *Haw. Handb.*126 (1976), Bayer

H. cv. *kewensis*

This species was described from a plant sent to von Poellnitz from Kew under the name *H. peacockii*, from which it was seen to differ markedly. Although von Poellnitz's setting up of a specific name is completely unjustified, since the wild origins of the plant were not even hinted at, it is of distinct appearance and widely in cultivation. It is therefore retained as a cultivar: *H.* cv. *kewensis* von Poelln.

It forms a stem in the manner of the section Coarctatae, with thick, triangular leaves of blackish-green, covered with concolorous tubercles, thickly on the backs of the leaves, less so on the upper surfaces. Leaves are about 3cm long and half as wide, with teeth on the margins.

H. kingiana von Poelln., *Fedde's Repert. Spec. Nov.* 41:203 (1937); & 44:218 (1938); *Haw. Handb.* 126 (1976), Bayer
Section Margaritiferae

After an interlude as a variety of *H. subfasciata*, a name which has been in question for many years, this species has regained its species status and is well-known in cultivation at present. It has been placed alongside the select large-growing band in the subgenus *Robustipedunculares*, with *H. pumila*, to which it bears some resemblance. It has wide, triangular leaves marked to a greater or lesser degree with tubercles, but not so prominently as in the larger-growing *H. pumila*, and less prominent and less brightly white than in that species. Its leaf colour differs too, in being a shining, light green, tending to become orange-red, especially towards the leaf tips, when grown in full light. It forms individual rosettes up to about 8cm tall and about 7cm wide, clumping slowly from the base after some years. It is not difficult to grow, although room for its strong root growth should be given. It will colour as indicated without losing its leaf-tips if given full light and sufficient water to keep it growing strongly.

It grows at Great Brak, Mossel Bay.

H. koelmaniorum Obermeyer & Hardy, *Fl. Pl. Afr.* 38:t.1502 (1967); *Cact. Amer.* 40:92 (1968), Hardy; *Haw. Handb.* 127 (1976), Bayer
Section Venosae

This close relative of *H. limifolia* represents the most northern Haworthia found, coming from Groblersdal in the Transvaal. It remains for the present little

H. kingiana

H. koelmaniorum

H. kingiana (glabrous form)

H. cv. *kuentzii*

known in cultivation. The photograph is of a plant in cultivation in the Karoo Botanic Garden, and there are other single plants in odd collections. Until 1980 little seemed to have been done to propagate this species but seedlings have appeared in the last year or two, and it is to be hoped that it will become more widely available.

It makes solitary rosettes, dark green becoming bronzed in full light. Leaves are about 7cm long, 2cm wide at the base, tuberculate with the tubercles in irregular rows right to the edge of the leaves.

The exact locality was not reported to prevent overcollecting.

H. krausiana and **H. krausii**—both Hort. Haage & Schmidt—are garden hybrids, of no appeal or significance.

H. cv. kuentzii Hort.

The origins of this hybrid are unknown to me, but it is of distinct appearance and is widely cultivated. It has affinities to *H. attenuata*, and has the same growth habit, but with shorter leaves, tending to become shortly columnar in time. The bright white tubercles make it an attractive collector's plant.

H. laetevirens Haw., *Suppl. Pl. Succ.* 43 (1819); *Monogr.* 9:5 & 10:3 (1836–63), Salm-Dyck; *Haw. Handb.* 12 (1976), Bayer

This is a small, light green species with lightly recurving leaves, referred to *H. turgida*.

H. lateganiae von Poelln., *Des. Pl. Life* 103 (1937); *Fedde's Repert. Spec. Nov.* 99 (1938), von Poellnitz; *Cact. Amer.* 15:176 (1943), J. R. Brown; *The First Fifty Haw.* 37 (1970) & *The Second Fifty Haw.* 36 (1975), Pilbeam; *Haw. Handb.* 12 (1976), Bayer

A longer-leaved, larger-growing variety of *H. starkiana*.

H. leightonii G. G. Smith (NOT 'leightoniae'), *Journ. S.A. Bot.* 16:10 (1950); *Cact. Amer.* 39:46 (1967), J. R. Brown; *Haw. Handb.* 12 (1976), Bayer

Bayer has reduced this species to varietal status under *H. cooperi*.

H. lepida G. G. Smith, *Journ. S.A. Bot.* 10:21 (1944); *The Second Fifty Haw.* 36 (1975), Pilbeam; *Haw. Handb.* 12 (1976), Bayer

Bayer reduces this species to synonymy with *H. cymbiformis*.

H. limifolia Marl., *Trans. Roy. Soc. S.Afr.* 409 (1908); *Cact. Amer.* 13:3 (1941), J. R. Brown; *The First Fifty Haw.* 21 (1970), Pilbeam; *Haw. Handb.* 12 (1976), Bayer

Section Venosae

The plethora of varieties and forms ascribed to this species is largely the work of Dr Flavio Resende of Portugal in the early 1940s. They were in the main not

H. limifolia var. *limifolia* fa. *limifolia*

based on authenticated field-collected plants, nor on field study. They are poorly illustrated in an obscure work in Portuguese, and must largely be discounted as of little significance. Bayer merely lists them in an Appendix in his 1976 Handbook.

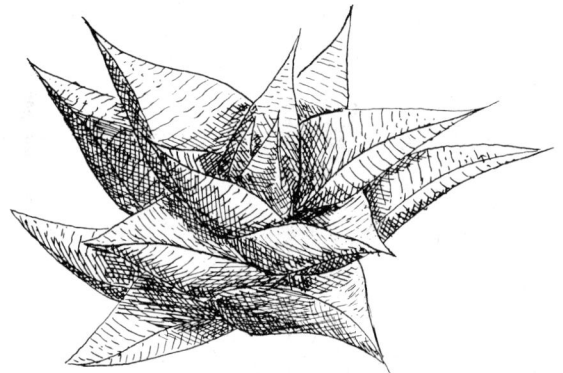

H. limifolia var. *limifolia* fa. *limifolia*

H. limifolia var. *limifolia* fa. *major*

To consider them one by one:

H. limifolia var. *limifolia*, the type, is well-known in cultivation, and has until recent years been by far the most common in collections. It makes a compact

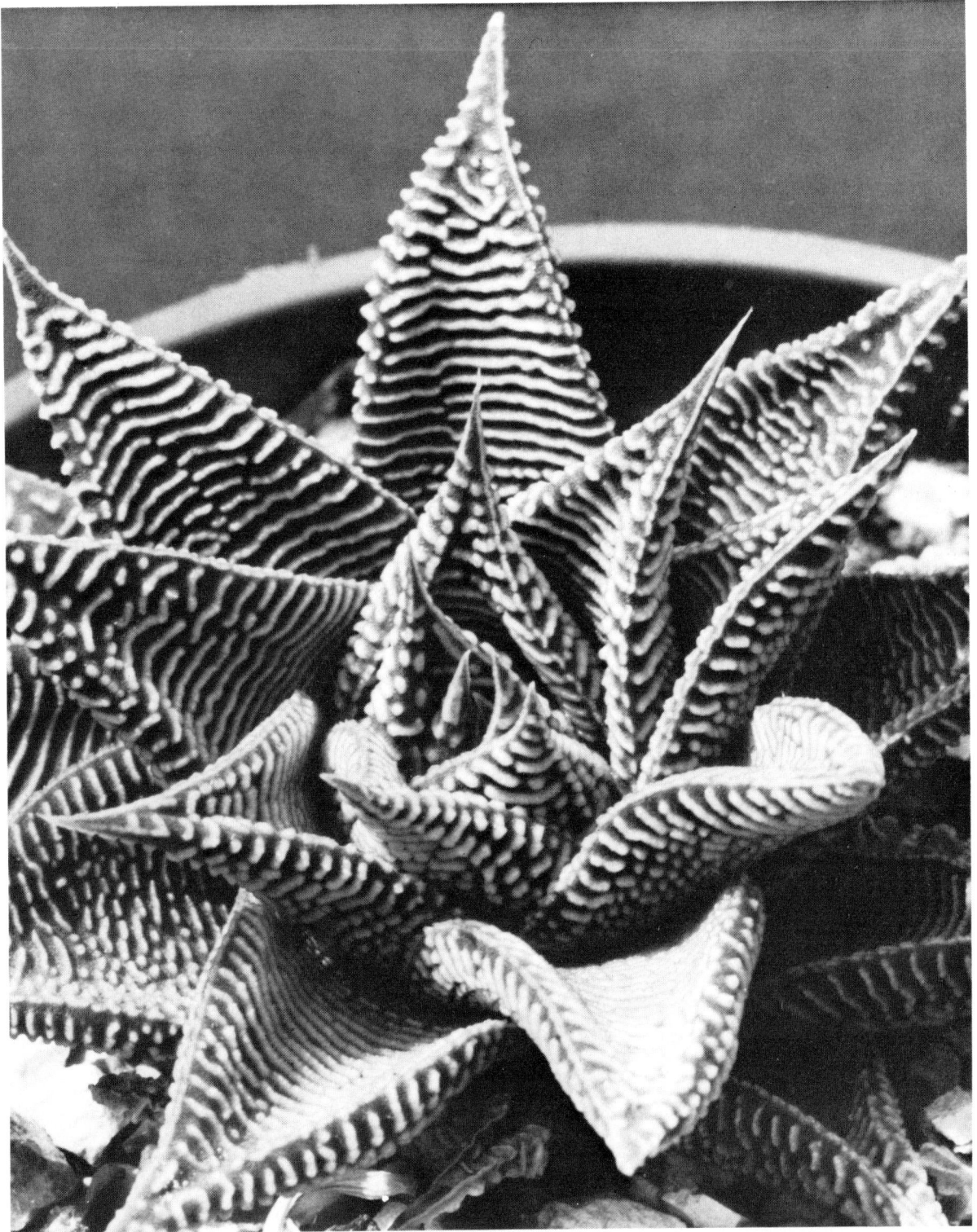

H. limifolia var. *striata*

rosette with leaves very slowly lengthening as they spread downwards from the growing-point in a tight spiral, so that the effect is of a tightly screwed-up plant. The rosettes are 8 to 10cm in diameter. The leaves are, inclined at a shallow angle, the lower nearly horizontal, and get to 5cm long (or longer in cultivation) and about 2cm broad at the base, broadly ovate-triangular in shape, tapering to a sharp point, the upper surface more or less concave, the back surface strongly convex with an obtuse keel, both surfaces with 15 to 20 transverse ridges the same colour as the leaf, which is generally dark green, becoming dark greenish-brown in full light. Suckers give rise to offsets around the plant often up to 10cm or more distant from the main rosette. Reported from the Barberton district in the Transvaal.

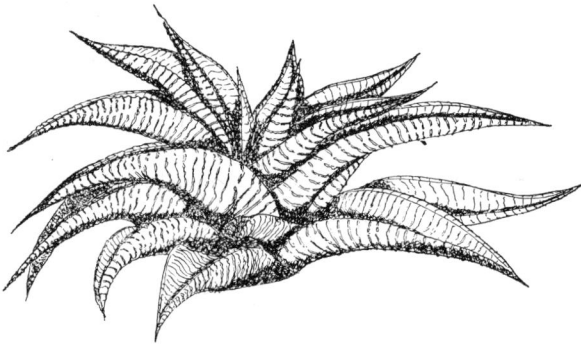

H. limifolia var. *limifolia* fa. *major*

H. limifolia var. *gigantea* Bayer, *Journ. S.A. Bot.* 28:215 (1962); *Haw. Handb.* 12 (1976), Bayer. This is a potentially massive variety, reported to have rosettes up to 23cm in diameter. The leaves are less distinctly ridged than the type, with an all-over roughening of the surface rather than ridges, making the plants less

H. limifolia var. *gigantea*

shining. It offsets stoloniferously, and has the same characteristic whorl of leaves as the type. In appearance it is perhaps the closest of this species to *H. koelmaniorum*. It is maintained as a good variety. Reported from Zululand, Natal.

H. limifolia var. *keithii* G. G. Smith, *Journ. S.A. Bot.* 16:4 (1950); *Cact. Amer.* 37:21 (1965), J. R. Brown; *Haw. Handb.* 126 (1976), Bayer. This variety has neither the prominent ridges nor the dark colouring of most other varieties, being an altogether lighter green, reddening in full light, and having barely raised ridges of tubercles somewhat broken, and numbering only 4 to 8 on each leaf. It is not upheld. Reported from Swaziland, in the Ubombo mountains among rocks at about 600m altitude.

H. limifolia var. *schuldtiana* Res., *Mem. Soc. Brot.* 3:92 (1943). I have grown a plant of this variety for some time, which originated from Resende. It seems to me to be intermediate between *H. limifolia* and *H. attenuata*, having the surface texture of the former, with less definitely defined ridges, but with the overall shape of leaf and rosette of the latter. It produces offsets in the manner of the latter too, not stoloniferously as with all other varieties of *H. limifolia*. The flower too owes more to *H. attenuata*. With its doubtful origins it must be discounted as probably of garden origin.

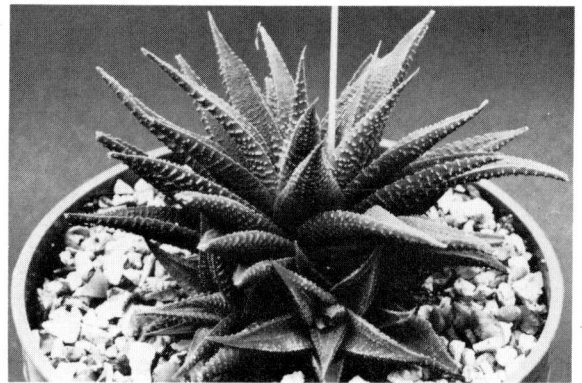

H. limifolia var. *schuldtiana*

H. limifolia var. *stolonifera* Res., *Mem. Soc. Brot.* 2:94 (1943); *Cact. Amer.* 19:39 (1947), J. R. Brown. This variety was differentiated by its looser-leaved habit and a facility for producing stoloniferous offsets freely. It should be pointed out that most, if not all, members of this species (except *H. limifolia* var. *schuldtiana*, for the reasons postulated) have the habit to a greater or lesser degree of producing offsets from stolons, and there is some variation in the arrangement of the leaves in the type, sufficient for one to discount the erection of this variety.

H. limifolia var. *stolonifera* fa. *major* Res., *Mem. Soc. Brot.* 2:94 (1943); *Cact. Amer* 34:46 (1962), J. R.

Brown. This is an altogether larger growing form than the type, with a looser arrangement of leaves, regularly tapering, and with ridges entire across the width of the leaves. It is retained as a distinctive collector's piece, and is widespread in cultivation. Reported from Blue Jay Ranch.

H. limifolia var. *stolonifera* fa. *pimentelii* Res., *Mem. Soc. Brot.* 2:94 (1943). It is apparent that the only difference of this form from the type is the more freely stoloniferous habit. In all other respects it equates to the type, and it is here regarded as synonymous.

H. limifolia var. *striata* var. nov. Pilbeam. This very distinctive variety was first pictured by Bayer in his paper setting up *H. limifolia* var. *gigantea* in 1962, and at that time its status was considered insufficiently determined. It has since become apparent that it is a discrete variety of this species with the beautiful characteristic of ridges of white tubercles instead of the concolorous ridges of most others in the species. It is therefore upheld here as a new variety as follows: *a typo foliis albo striatis differt*; differs from the type with its white striated leaves. Reported from an area north of Hluhluwe and, with shorter leaves, from the Mozaan river area, sometimes confused with var. *keithii*.

H. limifolia var. *ubomboensis* (Verd.) G. G. Smith, *Journ. S.A. Bot.* 16:4 (1950); *Fl. Pl. Afr.* t.818 (1941), Verd.; *Cact. Amer.* 25:11 (1953), J. R. Brown; *The Second Fifty Haw.* 38 (1975), Pilbeam; *Haw. Handb.* 164 (1976), Bayer. Dr Peter Brandham of the Jodrell Laboratory, Royal Botanic Gardens, Kew, has examined this variety cytologically, and categorically states that it is not a variety of *H. limifolia* but a species in its own right. It is therefore returned to its former standing, see under *H. ubomboensis*.

The species resolves as follows:

H. limifolia Marl var. *limifolia*.

H. limifolia var. *limifolia* fa. *major* (Res.) Pilbeam (syn. *H. limifolia* var. *stolonifera* var. *major* Res.).

H. limifolia var. *gigantea* Bayer.

H. limifolia var. *striata* Pilbeam.

H. lisbonensis Res., *Port. Acta Biol.* (B) 2:175 (1946); *Cact. Amer.* 25:139 (1953), J. R. Brown; *The First Fifty Haw.* 21 (1970), Pilbeam; *Haw. Handb.* 177 (1976), Bayer

This species was described by Resende from plants received from the Lisbon Botanic Garden, with affinities to *H. tortuosa* and *H. rigida*, themselves doubtful species to say the least. Its origin in the field is doubted, and it is discarded as probably of garden origin.

H. lockwoodii Arch., *Fl. Pl. Afr.* 20:t.792 (1940); *Ashingtonia* 1:59 (1974), Pilbeam; *The Second Fifty Haw.* 40 (1975), Pilbeam; *Haw. Handb.* 129 (1976), Bayer

Section Arachnoideae subsection Limpidae

This is indeed a rare species, with onion-like qualities, in having many thin leaves growing in a compact, ball-shaped rosette, and tending to dry back from the tips and enclose the inner part of the rossette in a paper-like sheath. The leaves are about 6cm long and 2 to 3cm wide, pointed at the apex, but almost round in outline, with very fine, small teeth on the margins; there is too an almost indiscernible keel. It is one of the few Haworthias I have lost, through overwatering I am sure, although it grew well for over a year for me. Perhaps my careful avoidance of water between the leaves at first was undermined by its apparent vigour. Certainly its collapse into a ball of pulp was preceded by a carefree period of overhead watering when I felt that it had accepted wholeheartedly its new, more kind situation. I was too kind; with this species I think you probably have to be cruel with the water to be kind. Reported from the Laingsburg area. Its nearest relative seems to be *H. decipiens*.

H. lockwoodii

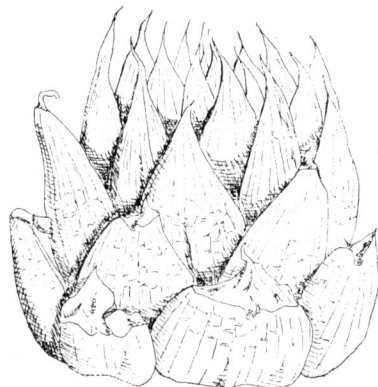

H. lockwoodii (in growth)

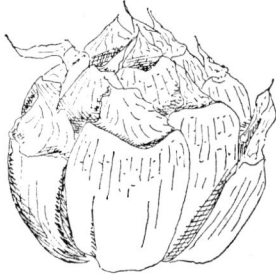

H. lockwoodii (at rest)

H. longebracteata G. G. Smith, NOT '*longi-bracteata*', a misspelling, *Journ. S.A. Bot.* 11:75 (1945); *Cact. Amer.* 25:73 (1953), J. R. Brown; *Haw. Handb.* 130 (1976) Bayer; *Excelsa* 5:86 (1975), Bayer

Reduced by Bayer to synonymy with *H. retusa*, this represents a narrower-leaved variant with lighter green leaves, turning orange-red in full sun. It is maintained at form level under *H. retusa*.

H. longiana von Poelln., *Fedde's Repert. Spec. Nov.* 41:203 (1937); & 44:213 (1938); *Des. Pl. Life* 9:78 (1937), von Poellnitz; *Cact. Journ.* 5:31 (1936), & 6:19 (1937), von Poellnitz; *Cact. Amer.* 16:179 (1944), J. R. Brown; *The First Fifty Haw.* 22 (1970), Pilbeam; *Cact. Succ. Journ. NSW.* 8:7 (1973) Bayer; *Haw. Handb.* 130 (1976), Bayer
Section Coarctatae subsection Attenuatae

This is, as you might be led to think from the name, a very long-leaved species, but it was named for Maj. F. R. Long, Superintendent of Parks at Port Elizabeth, not for its habit. The leaves are up to 25cm long, more

H. longiana

often about 15cm in cultivation, more upright than similar species, and the leaf-tips have an even stronger inclination to die back from the tips than in *H. attenuata* or *H. glabrata*. Sufficient water at the roots coupled with a well-drained growing medium is necessary to minimize this quite normal method of water-conservation by this species. The type has almost smooth leaves, shiny dark green, minutely tuberculate on the backs of the leaves in places, sometimes coalescing into indistinct longitudinal lines, but concolorous with the leaf. It clusters from the base to form tangled clumps. The variety, *H. longiana* var. *albinota* G. G. Smith, *Journ. S.A. Bot.* 14:44 (1948), was set up for a variant with more distinct and whitish tubercles, but the type is in any case more or less tubercled, and this character does not justify recognition at varietal level.

Reported from Cape Province, at Humansdorp, Hankey Road.

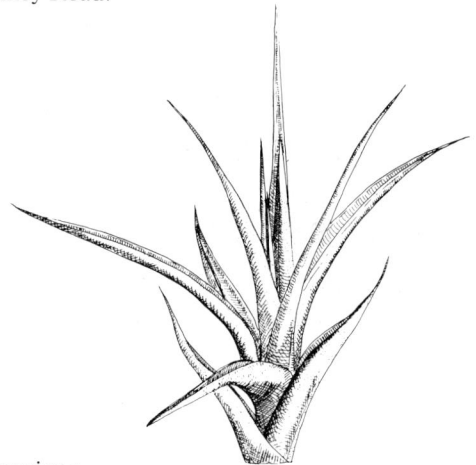

H. longiana

H. luteorosea Uitew., *Cactus. Vetpl.* 5:88 (1939); *Cact. Amer.* 37:127 (1965), J. R. Brown; *Cact. Succ. Journ. NSW.* 7:9 (1971) & 8:103 (1972), Bayer; *Nat. Cact. Succ. Journ.* 27:51 (1972), Bayer; *The Second Fifty Haw.* 6 (1975), Pilbeam; *Haw. Handb.* 130 (1976), Bayer

This species is merely an attractively pink-flowered form of *H. herbacea*, having all the characteristics of that species and indistinguishable except when in flower. The flower-colour of *H. herbacea* varies from pale pink to that in this form, which is a yellowish salmon-pink. For the sake of the flower-colour it is worth looking out for, since the flowers of this genus are not in general outstanding, but the feature does not warrant recognition at any level.

H. maculata (von Poelln.) Bayer, *Fedde's Repert. Spec. Nov.* 49:25 (1940); *Cact. Succ. Journ. NSW.* 8:9 (1971) Bayer; *The Second Fifty Haw.* 8 (1975), Pilbeam; *Haw. Handb.* 130 (1976), Bayer
Section Retusae subsection Turgidae

Formerly a variety of *H. schuldtiana*, described by von Poellnitz, this now has species status according to Bayer, whose fieldwork has established that it is a discrete population quite apart from the former *H. schuldtiana*. It occurs in an area south of Worcester at the Brandvlei dam, and has affinities with *H. herbacea* and *H. reticulata*. Because of the character of the flower and the flowering time, among other reasons, Bayer regards it as a separate element from these, warranting specific recognition. The leaves are rounded on the upper surface and recurve gradually, the margins and keel are toothed as are, to a lesser extent, the leaf surfaces, and where the teeth arise there is usually a translucent fleck in the surface, giving the leaf an appearance of being water filled. In full light it will colour beautifully to almost violet.

H. maculata

H. magnifica von Poelln., *Fedde's Repert. Spec. Nov.* 33:240 (1933); & 43:104 (1938); *Haw. Handb.* 131 (1976) Bayer; *Nat. Cact. Succ. Journ.* 32:18 (1977), Bayer
Section Retusae subsection Retusae

Bayer has relegated several species to a position beneath this species (*H. schuldtiana, H. sublimpidula, H. maraisii, H. notabilis, H. atrofusca* and *H. paradoxa*) as well as deposing the several previously erected varieties of *H. schuldtiana* to synonymy, except for *H. schuldtiana* var. *major*, which he maintains at varietal level. The photographs illustrate the variability, and included is one of the newly established variety by Bayer, *H. magnifica* var. *meiringii*, from Bonnievale. The species resolves as follows:

H. magnifica von Poelln. var. *magnifica*.

H. magnifica var. *atrofusca* (G. G. Smith) Bayer (syn. *H. atrofusca*).

H. magnifica var. *major* (G. G. Smith) Bayer (syn. *H. schuldtiana* var. *major*).

H. magnifica var. *maraisii* (von Poelln.) Bayer (syn. *H. maraisii, H. schuldtiana* and *H. sublimpidula*).

H. magnifica var. *meiringii* Bayer.

H. magnifica var. *notabilis* (von Poelln.) Bayer (syn. *H. notabilis*).

H. magnifica var. *paradoxa* (von Poelln.) Bayer (syn. *H. paradoxa*).

The type, *H. magnifica* var. *magnifica*, occurs southeast of Riversdale and west of Heidelberg. Bayer comments that it is not easy to rationalize the distinction between this, the type, and *H. magnifica* var. *maraisii*, except that the type is larger, less scabrous, with leaves very acuminate and with a longer end-bristle; the flowers too are larger and have a long tube. The type has been little known since its erection in 1933, and no photograph was published at this time, nor since then except in the last few years by Bayer. A masquerader has been commonly seen under this name in collections, and this larger-leaved plant is near to Bayer's *H. retusa* var. *acuminata*. The main apparent difference between the type and other varieties is the much smoother end-area of the leaf.

H. magnifica var. *magnifica*

H. magnifica var. *atrofusca* (G. G. Smith) Bayer, *Nat. Cact. Journ.* 32:18 (1977), Bayer; *Journ. S.A. Bot.* 14:41 (1948), G. G. Smith; *Haw. Handb.* 100 (1976), Bayer. This variety is a small gem in the genus. It grows slowly and colours well in full light to dark blackish green. The upper surface of the small leaves, barely 1cm wide or less at their widest point, is roughened from a covering of small tubercles the same colour as the leaf and translucent, so that when it is well-coloured from exposure to sunlight the end-area has a black gem-like quality. There are 3 to 5 indistinct, longitudinal lines in this upper portion, the centre one longer but reaching only halfway to the tip of the leaf. The keels and margins have small, transparent teeth, longer towards the tip of the leaf, but hardly noticeable, and absent on the keel except at the tip. The end-area in this variety is somewhat rounded. In strong light it will remain more compact and shorter-leaved, and it remains solitary for some time, only sparingly offsetting from the base on reaching its full size of rosette of about 6 or 7cm in diameter. I prefer to see it grown a little 'hard', in a good amount of sunlight with sufficient water to keep it growing strongly. It makes thick roots which require room to develop in order to serve the plant well, but an open gritty compost is best for this variety. Reported from the Riversdale Division in Cape Province.

H. magnifica var. *atrofusca*

H. magnifica var. *major* (G. G. Smith) Bayer, *Nat. Cact. Succ. Journ.* 32:18 (1977); *Haw. Handb.* 132 (1976), Bayer; *Journ. S.A. Bot.* 12:1 (1946), G. G. Smith as *H. schuldtiana* var. *major*. This variety is a collector's beauty, an overall larger variety than the type, with more leaves to each rosette and a tendency to stay solitary for some time. Apart from these distinctions the end-areas of the leaves have many tubercles armed with small bristles. In cultivation, if given good light conditions, this variety stays flatter, with the end-areas forming a tightly-fitting, marvellously colourful mosaic, reminiscent in its colouring of *H. emelyae*, with which it is continuous, according to Bayer, via an intermediate population at Sandkraal, a few miles east of the locality for this variety, which is north of Garcias Pass at Muiskraal, where it grows amongst sandstone.

H. magnifica var. *major*

H. magnifica var. *maraisii* (von Poelln.) Bayer, *Haw. Handb.* 132 (1976); *Nat. Cact. Succ. Journ.* 32:18 (1977); *Fedde's Repert. Spec. Nov.* 38:194 (1935), von

Poellnitz—as *H. maraisii*, makes clusters of small rosettes, each about 8cm in diameter. The end-area is usually tuberculate, and the margins set with prominent teeth, although the amount of tuberculation and bristling varies a great deal. It has the tendency, as similar species do, to colour to a shining, blackish green in full light, with sufficient water to keep it growing well, giving it a most attractive appearance. It comes from Stormsvlei near Riviersonderend, where it is now scarce, but spreads widely varying in appearance over a wide area. Bayer points out particularly the small, very green flowers appearing in habitat in March and April, compared with the flowers of *H. mirabilis*, which flowers earlier, with brownish flowers and with buds longer and markedly narrow and biarcuate (S-shaped); there is one population south of Stormsvlei which appears to be distinctly hybrid between this variety and *H. mirabilis*.

H. magnifica var. *maraisii*

H. magnifica var. *meiringii*

H. magnifica var. *meiringii* Bayer, *Haw. Handb.* 134 (1976); *Nat. Cact. Succ. Journ.* 32:18 (1977), is a variety set up by Bayer in his Handbook to embrace a population at Bonnievale, which superficially resembles *H. herbacea*. In cultivation this variety tends to stay flatter and grow much more slowly than *H. herbacea*, while the colouring has no tendency to be of a chlorotic appearance as is often the case with *H. herbacea*. But it is distinctly different from other varieties of *H. magnifica* in its erect leaves with little indication of a distinct end-area, and its larger bristles which freely adorn the leaves.

H. magnifica var. *notabilis* (von Poelln.) Bayer, *Haw. Handb.* 141 (1976); *Nat. Cact. Succ. Journ.* 32:18 (1977); *Fedde's Repert. Spec. Nov.* 44:134 (1938) von Poellnitz—*H. notabilis*. This variety has been vexatious since its original erection as a species in 1938. Its relationship is not made easier to determine because of its proximity to the complex species *H. herbacea* and *H. reticulata*. But Bayer, after extensive field study, concludes it to be an ecotype of *H. magnifica* and reduces it accordingly. It occurs in the southern foothills of the Langeberg mountains at Robertson. The flower is characteristically that of *H. magnifica*, except that it tends to be yellow-throated instead of green. The type locality is Wolfkloof, Robertson, but it occurs from Klaasvoogds to Buitenstekloof. In form it differs considerably from the type in having less markedly separate end-areas to the leaves, without the sudden angle at this point of the leaf, the outline being more of a curve than a sharp angle. The end-area is less tuberculate than the type and leans more towards the general appearance of *H. herbacea*, but altogether a different colouring from the latter, becoming purplish-brown in full light compared with the yellowish colouring of an exposed *H. herbacea*.

H. magnifica var. *notabilis*

H. magnifica var. *paradoxa* (von Poelln.) Bayer, *Haw. Handb.* 143 (1976); *Nat. Cact. Succ. Journ.* 32:18 (1977); *Fedde's Repert. Spec. Nov.* 33:239 (1933), von Poellnitz—*H. paradoxa*. This is the largest-leaved of the species except for the type, and has the large sub-erect leaves more typical of *H. mirabilis*, but the flowering time and the flower colour have led Bayer to ascribe it to its present place with *H. magnifica*. It

occurs near Vermaaklikheid. It is a beautiful dark green variety, with somewhat smoother end-areas to the leaves than most other varieties, but with irregular longitudinal lines of tubercles on this part of the upper surface and light coloured tubercles and translucent flecks on the lower surface, towards the apex. It should be grown in fairly strong light to maintain the closeness of the leaves in a tight rosette, which if kept too shaded will flop open and allow the leaves to separate. As with others of this species it looks best if allowed to cluster into a low mound of rosettes, which it will do readily.

H. magnifica var. *paradoxa*

H. magnifica var. *paradoxa*

H. cv. mantelii Uitew., *Succulenta* 37 (1947); *The First Fifty Haw.* 23 (1970), Pilbeam

The only named hybrid of *H. truncata*, although there are many, and some much more attractive in cultivation. It was described as a hybrid between *H. truncata* and *H. cuspidata*, the latter considered itself to be a hybrid, so that such crosses could result in a variety of different-looking plants. Some of the best *H. truncata* crosses are with species of the Section Retusae, but it has been successfully hybridized with a number of other species, including *H. limifolia*, *H. arachnoidea* and *H. venosa*. None really warrants more than curiosity value.

H. cv. *mantelii*

H. cv. *mantelii*

H. maraisii von Poelln., *Fedde's Repert. Spec. Nov.* 38:194 (1935); & 43:104 (1938); *Cact. Amer.* 12:175 (1940), J. R. Brown; *Haw. Handb.* 132 (1976), Bayer as amended in *Nat. Cact. Succ. Journ.* 32:18 (1977), Bayer.

Referred to *H. magnifica*.

H. margaritifera (L.) Haw., *Sp. Pl.* 322 (1753), Linnaeus, as *Aloe pumila* var. *margaritifera*; *Suppl. Pl. Succ.* 55 (1819), Haworth; *Monogr.* 6:5 (1836–63), Salm-Dyck; *Des. Pl. Life* 6:119 (1934), J. R. Brown;

Cact. Succ. Journ. NSW. 8:9 (1971), Bayer; *Haw. Handb.* 133 (1976), Bayer

There is no doubt that C. L. Scott is right in asserting that the first-used name for this well-known species is *H. pumila*. See under that name.

H. marginata (Lam.) Stearn, *Encycl.* 89 (1783), Lamarck; *Cact. GB* 7:39 (1938) Stearn; *Monogr.* 5:1 (1836–63), Salm-Dyck; *Cact. Amer.* 19:119 (1947), J. R. Brown; *The Second Fifty Haw.* 40 (1975), Pilbeam; *Haw. Handb.* 133 (1976)
Section Margaritiferae

H. marginata

This species occurs close to *H. pumila* (syn. *H. margaritifera*) in habitat, and the two species are known to hybridize. The name *H. albicans*, a synonym, referred to the translucently white epidermis of this species, and the correct prior epithet, *H. marginata*, to the thickened, whitish, cartilaginous leaf-margins and keels. The leaves are stiff and Agave-like with a horny tip. The margins of the leaves vary considerably from a continuous, barely discernible narrow edging to prominent, thick ridges bleeding into individual tubercles near to the margins on the flat surfaces of the leaves. The more tuberculate forms (possibly where the species meets *H. pumila*) gave rise to the species *H. uitewaaliana*, which Bayer has discarded as synonymous. He also discards the former varieties: *H. marginata* var. *laevis* (Haw.) Jacobs., *H. marginata* var. *ramifera* (Haw.) Jacobs. and *H. marginata* var. *virescens* (Haw.) Uitew., as being merely variants of the species, not warranting recognition at any level. Certainly little real idea can be gained of what was originally described, and with the variability of the species it is hardly worth the effort of trying to determine. An adequate-sized pot is essential to grow this species well, and in the early years regular repotting each year to accommodate the hungry roots is necessary for best results.

H. marginata is found in Bredasdorp, the upper Breede River Valley and the Robertson Karoo area.

H. marginata

H. marginata from North Ashton

H. marginata × *H. minima*

H. maughanii

H. marumiana Uitew., *Cact. Vetpl.* 33 (1940); *Cact. GB* 9:20 (1947), Uitewaal; *Cact. Amer.* 33:48 (1961), J. R. Brown; *Cact. Succ. Journ. NSW.* 8:127 (1972), Bayer; *Haw. Handb.* 133 (1976), Bayer
Section Arachnoideae subsection Proliferae

   This small species clumps readily and grows vigorously into a low mound of small rosettes, each up to about 5cm across, colouring dark purple-brown in full light. There is considerable variation of form in the amount of bristles or teeth, sometimes confined to the margins and keel, sometimes on the upper and lower surfaces as well, and sometimes nearly absent or very small. It varies too in the size and narrowness of the leaf and in the amount of translucent flecking in the upper part of the leaf, usually related to the presence or absence of bristles, as these normally arise from these flecks. Although the species was reported originally from Ladismith and Mossel Bay, Bruce Bayer has spent some considerable time in these areas fruitlessly searching for it. He finally located it in the Tarkastad area, and there are records of similar plants at Cradock and at Beaufort West, which indicates wide distribu-

tion. There is a hybrid swarm between this species and *H. cymbiformis* at a near-inaccessible spot at Klipplaat Dam, near Whittlesea, north of Cathcart.

   As regards cultivation it presents no problems at all, and is a ready candidate for Gordon Rowley's 'Haworthia balls' treatment: when clumps get big enough Gordon suggests tipping two out of their pots, placing the soil masses together and tying the two together back-to-back to hang up as a sort of 'hanging basket without the basket'. I cannot bring myself to treat these lovely plants in this way, but it serves to illustrate that this species will tolerate almost any treatment. It looks best if kept in fairly strong light to colour the leaves, without shrivelling them too much. It will take a good amount of water in the summer, but this may induce more lush, green growth; a happy medium should be aimed at.

H. maughanii von Poelln., *Fedde's Repert. Spec. Nov.* 31:85 (1932); & 41:205 (1941); *Des. Pl. Life* 9:42 (1937), J. R. Brown; & 16:12 (1944), J. R. Brown; *Cact. Amer.* 20:99 (1948), P. C. Hutchinson; & 27:67 (1955), J. R. Brown; *The First Fifty Haw.* 23 (1970), Pilbeam; *Haw. Handb.* 134 (1976), Bayer
Section Fenestratae

This wonderful species is well-known to collectors, with truncate, rounded ends to the finger-like leaves, which in habitat lie mainly below ground-level, with just the truncated, translucent tips of the leaves showing at ground level. Unlike its companion species in the section (*H. truncata*) it tends to remain solitary, although in cultivation if treated well it will in time offset. It differs from this species in having more rounded, translucent end-areas to the leaves, which are arranged in a spiral, compared with a distichous (fan-like) arrangement in *H. truncata*. It is not difficult to grow, but will do its best for you if enough room is given to accommodate its thick, long, contractile roots, which it persists in sending down in the mistaken belief that it may have to haul itself down to avoid the worst effects of the blistering, South African sun. Watering should not be heavy, though provided enough light and drainage is given it will take a fair amount while growing strongly. Propagation is not difficult from seed, and this is much the best way with such a few-leaved species, although, if you can bear to do so, leaves may be removed and rooted, and will send up several plantlets each. Unfortunately the outside leaves may well be already too senile to undertake this process; best results are obtained from leaves still turgid and full of life, which this species can ill afford to have removed.

It occurs in the Little Karoo in the Oudtshoorn and Calitzdorp area in conjunction with *H. truncata*, and in a separate area alone, west of their common occurrence.

H. marumiana

H. mclarenii von Poelln., *Des. Pl. Life* 11:107 (1939)
The identity of this species has been in doubt since its description. Bayer relates it to *H. aristata* as it occurs at Barrydale.

H. minima (Ait.) Haw., *Syn. Pl. Succ.* 92 (1812); *Hort. Kew.* 1:468 (1789), Aiton; *Monogr.* 6:6

H. maughanii

(1836–63), Salm-Dyck; *Cact. Amer.* 29:22 (1955), J. R. Brown; & 43:157 (1971), Bayer; *The Second Fifty Haw.* 42 (1975), Pilbeam; *Haw. Handb.* 135 (1976), Bayer
Section Margaritiferae

For many years this has been a common misnomer for *H. translucens* subsp. *tenera*, a small, clumping, grey-green, bristly species. But the name was invalid in this application as it had already been used for a species closely related to *H. pumila* (syn. *H. margaritifera*), but smaller in all ways. The matter was corrected by von

Poellnitz as regards the misnomer, and recently Bayer in his revision of the genus reinstated *H. minima* as a species and placed it in the subgenus Robustipedunculares. It is in some of its forms the most attractive of the subgenus, heavily marked with prominent, white tubercles all over the leaves, so dense sometimes as to give the plant a frosted appearance. It comes from a wide area, from the coastal area between Bredasdorp and Mossel Bay inland to Riversdale, and, as might be expected with such a wide distribution, the form varies considerably from densely tuberculate to more sparsely tuberculate and less white tubercles, to, rarely, forms with no tubercles at all. The flower is characteristic of this subgenus, thick-stemmed and branching strongly, with flower-tubes abruptly joined to the pedicels, nearly regular flowers, developing into large, roundish seed-capsules and large, disc-like, dark brown seeds. Full light accentuates the tubercles in their whiteness, and a large enough pot to allow the roots to expand fully will allow the plant to grow quite rapidly and cluster to form a clump the size of a dinner plate.

H. minima (coastal form)

H. minima

H. minima

H. mirabilis Haw., *Syn. Pl. Succ.* 95 (1812); *Trans. Linn. Soc.* 7:9 (1804)—*Aloe mirabilis*; *Monogr.* 9:1 (1836–63), Salm-Dyck; *Des. Pl. Life* 8:66 (1936), J. R. Brown; *Cact. Amer.* 35:61 (1963) J. R. Brown; *Aloe* 11:8 (1973) C. L. Scott; & 12:89 (1974), Bayer; *The Second Fifty Haw.* 43 (1975) Pilbeam; *Haw. Handb.* 136 (1976), Bayer; *Excelsa* 7:37 (1977), Bayer
Section Retusae subsection Retusae

This is one of a group of most attractive species for the collector, which includes such other choice dark jade examples as *H. magnifica*, *H. mutica* and *H. heidelbergensis*. Some of the most popular former species have been united with *H. mirabilis*, i.e. *H. badia* and *H. mundula*. The shining, darkly transparent end-area of the leaves of the type is usually bisected almost to the tip by one prominent central line intruding from the opaque lower part of the leaf; two shorter lines or more protude from the base area, but not so prominently. Bayer has expanded the concept of this species vastly to include the two former species mentioned above as subspecies and to reduce to synonymy a great many others formerly embraced by *H. triebnerana*, under which von Poellnitz had described no fewer than 12 varieties on minor differences, almost none of which can now be identified with certainty. Also swept in is the beautiful *H. emelyae* var. *beukmanii* (but not *H. emelyae*), *H. rossouwii* and *H. willowmorensis*.

Undoubtedly it would be a tragedy for collectors to lose some of these names completely, since their distinctive form warrants some recognition: for the more outstanding and distinctive therefore the names are retained as forms, and the species resolves as follows:

H. mirabilis Haw. subsp. *mirabilis* fa. *mirabilis* (syn. *H. triebnerana, H. willowmorensis, H. rossouwii, H. nitidula*).

H. mirabilis subsp. *mirabilis* fa. *rubrodentata* stat. nov. (syn. *H. triebnerana* var. *rubrodentata* Triebn. & von Poelln., in *Fedde's Repert. Spec. Nov.* 47:10 (1939)), a distinctive form with longer, narrower, more erect leaves than the type, named for its reddish tubercles and marginal teeth, the latter notably reddening when exposed to full light. It is found west of Genadendal.

H. mirabilis subsp. *mirabilis* fa. *sublineata* stat. nov. (syn. *H. triebnerana* var. *sublineata* von Poelln., in *Fedde's Repert. Spec. Nov.* 44:135 (1938)), a small form with long, attenuated, prominently toothed leaves, which recurve in an arc rather than the angled recurve in the type, to make a most attractive, many-leaved rosette, and forming clusters in time. It comes from sandstone hills immediately south of Bredasdorp. *H. rossouwii* seems closest to this form.

H. mirabilis subsp. *mirabilis* fa. *rubrodentata*

H. mirabilis subsp. *mirabilis* fa. *mirabilis*

H. mirabilis subsp. *mirabilis* fa. *sublineata*

H. mirabilis subsp. *mirabilis* fa. *beukmannii*

H. mirabilis subsp. *mirabilis* fa. *napierensis*, stat. nov. (syn. *H. triebnerana* var. *napierensis* Triebn. & von Poelln., in *Fedde's Repert. Spec. Nov.* 47:11 (1939), and *H. triebnerana* var. *turgida* Triebn., in *Fedde's Repert. Spec. Nov.* 47:11 (1939), a form with thick, inflated leaves, suberect and notably translucent for a large part of the end-area. It comes from the farm Skietpad, 6 miles north of Napier.

H. mirabilis subsp. *mirabilis* fa. *beukmannii* stat. nov. (syn. *H. emelyae* var. *beukmannii* von Poelln., in *Fedde's Repert. Spec. Nov.* 49:29 (1940)), a distinctive, beautiful form, and the largest-leaved of the species, with translucent end-areas 3cm across or more. It comes from the farm Skuitsberg in the Caledon district, and Bayer states categorically that it has no association with *H. emelyae*.

H. mirabilis subsp. *mirabilis*

H. mirabilis subsp. *badia* (von Poelln.) Bayer (syn. *H. badia*), a stunningly beautiful subspecies with large, chestnut-coloured translucent-ended leaves, recurving strongly and growing slowly to form a rosette of about a dozen leaves. Bayer reports that it grows in a rather restricted area west of Napier, in pebbly, quartzitic soil associated with sandstone. The type of this species also grows in this area, but further north and in shale outcrops.

H. mirabilis subsp. *mirabilis* fa. *napierensis*

H. mirabilis subsp. *badia*

H. mirabilis subsp. *badia*

H. mirabilis subsp. *mundula* (G. G. Smith) Bayer (syn. *H. mundula*), a former species reduced by Bayer to subspecific rank under *H. mirabilis*. It represents the smaller end of the scale in this very variable species, and is distinguished by its short, thick, strongly recurving leaves, blunter than most others in the species. It occurs separately from *H. mirabilis* in a single known population south-west of Bredasdorp. The plant often seen in collections labelled *H. otzenii*, with short dark green leaves, and forming such large clusters so vigoriously that I doubt its parentage, seems nearer this subspecies than anything else.

H. mirabilis subsp. *mundula*

H. monticola (Bak.) Fourc., *Trans. Roy. Soc. S.Afr.* 21:78 (1932); *Haw. Handb.* 137 (1976), Bayer

Bayer equates this species to *H. angustifolia*. The name arose only because Fourcade considered Baker's description of *H. angustifolia* to differ from that species; he therefore renamed Baker's species as *H. monticola*. It is a name unrelated to any known plants and best forgotten.

H. morrisiae von Poelln., *Kakteenk.* 132 (1937); *Haw. Handb.* 137 (1976), Bayer

Bayer has reduced this species to varietal status under *H. scabra*.

H. mucronata Haw., *Suppl. Pl. Succ.* 50 (1819); *Fedde's Repert. Spec. Nov.* 45:168 & 169 (1938), von Poellnitz; *Haw Handb.* 137 (1976), Bayer

This is indeed a confused species, having been splintered into numerous varieties by von Poellnitz, mostly on the basis of minor variations, and often based on individual plants sent to him. There is no clear idea of the real identity of Haworth's original described species, and Bayer rightly discards the epithet as a source of confusion. *H. mucronata* var. *limpida* fa. *inconfluens* von Poelln., is taken as one of only two forms of the species which can 'clearly be related to field populations' and is upheld at varietal level beneath *H. habdomadis*; the other is the former *H. mucronata* var. *morrisiae* von Poe!!n., also upheld by Bayer as a variety of *H. habdomadis*. See under that species.

H. mundula G. G. Smith, *Journ. S.A. Bot.* 12:8 (1946); *Haw. Handb.* 139 (1976), Bayer

Bayer refers this species to subspecific status under *H. mirabilis*.

H. musculina G. G. Smith, *Journ. S.A. Bot.* 14:49 (1948); *Cact. Amer.* 33:46 (1961), J. R. Brown; *The First Fifty Haw.* 25 (1970), Pilbeam; *Nat. Cact. Succ. Journ.* 28:80 (1973), Bayer; *Haw. Handb.* 139 (1976), Bayer

This species is not upheld in Bayer's paper on *H. reinwardtii* and *H. coarctata* (*Nat. Cact.* ref above). Bayer asserts that it is not discontinuous with the range of forms of *H. coarctata*, and discards it accordingly.

H. mutabilis von Poelln., *Fedde's Repert. Spec. Nov.* 44:132 (1938); *Des. Pl. Life* 10:126 (1938), von Poellnitz; *Cact. Amer.* 41:57 (1969), J. R. Brown; & 43:157 (1971), Bayer; *Haw. Handb.* 139 (1976)

This species is disposed of by Bayer as on evidence of the finder, E. G. Payne, it was erected on the basis of an atypical specimen of *H. minima*. The name is therefore discarded.

H. mutica Haw., *Revis. Pl. Succ.* 55 (1821); *Monogr.* 9:3 (1836-63), Salm-Dyck; *Haw. Handb.* 139 (1976) Bayer; *Excelsa* 8:50 (1878), Bayer
Section Retusae subsection Retusae

This species has been lost to collectors for many

years, and its recent reintroduction represents indeed a worthwhile addition to any collection, since it is one of the most attractive in the genus. There is a marvellous painting at Kew in the Herbarium Library, over a hundred years old, but still full of colour and vitality as though it had been painted yesterday and were still barely dry. The species comes from the Bredasdorp to Riviersonderend area, and is close in appearance to *H. pygmaea*, yet distinct when known, with less roughened end-areas to the leaves, translucent with the appearance of ground glass, dark green colouring to pinkish-brown in full light, and generally with more leaves to the rosette. It offsets slowly, and presents no problems in cultivation, except that sufficient light should be given to keep it compact: the lower part of the leaves below the translucent end-area should rarely be visible at all.

H. mutica

H. nigra (Haw.) Bak., *Journ. Linn. Soc.* 18:203 (1880); *Phil. Mag.* 46:302 (1825), Haworth—*Apicra nigra*; *Haw. Handb.* 140 (1976), Bayer; *New Haw. Handb.* 70 (1982)
Section Trifariae subsection Caulescentes

As admirably written-up by Mrs P. Roberts-Reynecke of Cape Town University in a paper, as yet unpublished, on the genus *Astroloba* Uitew., this species was confused at the beginning of the 19th century with *Astroloba aspera* (Willd.) Uitew. (then known as *Apicra aspera* (Willd.), and the identity of the type of the species is in doubt. Any confusion in the reader's mind between these two is quickly resolved by looking firstly at the body colouring, which for *H. nigra* is dark green turning almost black in strong light conditions, while *A. aspera* is mid-green, with reddish tones in full light conditions. In addition *H. nigra* is clearly trifariate, i.e. with leaves arranged in three rows, straight or slowly spiralling, while *A. aspera* has leaves arranged in five rows and tightly spiralling to make them difficult to ascertain. And the flowers of course clinch the matter, being typical of each genus (see pp. 16, 147). Bayer discards all the varieties and forms, since the species is widely distributed and very

variable. Certainly most of them were based on minor differences of the leaf in length and shape, and at most could be regarded as forms of the species. And even at this level they are hardly worth retaining, except for the more notably different. I have retained only two forms apart from the type, to represent the shorter- and longer-leaved forms.

H. nigra fa. *nigra*

H. nigra fa. *nigra*

The species resolves as follows:

H. nigra fa. *nigra*, forming stems to about 8 or 10cm, with leaves in three distinct rows, more or less spiralling, leaves triangular, three-sided, the top of upper side concave, furrowed in the centre, the lower side strongly marked with tubercles the same colour as the leaf, the margins with a row of tubercles just inside the edge of the leaves. Offsets are formed by means of stolons, appearing from beneath the soil around the main stem, sometimes to several centimetres away. A wide pan is advisable for cultivation so that the stolons have room to come up around the main stem, otherwise in too narrow a pot they hit the sides still going down and will circle endlessly at the bottom of the pot sometimes finding their way out through the drainage holes. Reported from Preiska; Bulkraal; Somerset East; Ladismith; Aberdeen; Cradock; Willowmore; Steytlerville; Jansenville and Wellington, all in the Cape Province, although Bayer considers the last location to be 'incongruous'.

H. nigra fa. *angustata* (von Poelln.) Pilbeam, stat. nov. (syn. *H. nigra* var. *angustata* (von Poelln.) Uitew., *Succulenta* 51 (1948); and *H. schmidtiana* var. *angustata* von Poelln., *Kakt. u.a. Sukk.* 10:169 (1937); *Cact. Amer.* 17:105 (1945) & 19:25 (1947), J. R. Brown; *The First Fifty Haw.* 25 (1970), Pilbeam). This is a very attractive form with longer leaves than the type, recurving strongly; the length can be as much as 6 or 7cm. Reported particularly from Bulkraal near Slagtersnek; Ladismith; Prieska.

H. nigra fa. *angustata*

H. nigra fa. *angustata*

H. nigra fa. *nana* (von Poelln.) Pilbeam, comb. nov. (syn. *H. nigra* var. *diversifolia* fa. *nana* (von Poelln.) Uitew., *Succulenta* 51 (1948); and *H. schmidtiana* var. *diversifolia* fa. *nana* von Poelln., *Fedde's Repert. Spec. Nov.* 44:240 (1938); *Cact. Amer.* 17:165 (1945) J. R. Brown. This form represents the smallest of the species, with leaves no larger than 1.5cm, with just the leaf-tips recurving. In common with others in the species it clusters by stolons. Reported from Beaufort West.

H. nigra fa. *nana*

It should be emphasized that these last two described forms are extremes; there are intermediates between the type and both of them.

H. nitidula von Poelln., *Des. Pl. Life* 11:192 (1939); *Cact. Amer.* 18:89 (1946); & 26:73 (1953), J. R. Brown; & 52:10 (1980), Bayer; *Aloe* 11:26 (1973), Scott

Bayer equates this species with *H. mirabilis*, in a broad concept of a variable, widespread species.

H. nortieri G. G. Smith, *Journ. S.A. Bot.* 12:13 (1946); & 16:6 & 7 (1950)—var. *giftbergensis* & var. *montana* G. G. Smith; *Cact. Amer.* 32:27 (1960), J. R. Brown; *The First Fifty Haw.* 26 (1970), Pilbeam; *Haw. Handb.* 141 (1976), Bayer
Section Retusae subsection Turgidae

This is the species with the cachet of a yellow flower (or rather a yellow-throated flower). The leaves are finely bristled and have many translucent flecks, which; when the rosette has coloured in full light to an attractive purple-brown, show up well, especially with the light behind them. Rosettes are not large, about 5cm or so across, and they have a tendency in cultivation to rest for long periods after flowering, shrivelling considerably. Offsets are produced from around the base, sparingly. Leaves are narrow-triangular and taper steadily to a point with a long bristle, which often dries, but persists on the leaf. The two former varieties are little different, except in their overall size, both being smaller than the type; *H. nortieri* var. *giftbergensis* is found in the Giftberg mountains near Van Rhynsdorp, and *H. nortieri* var. *montana* from Pakhuis Pass, Clanwilliam Division. Bayer declares them synonymous with the type, but maintains *H. globosiflora* at varietal level under *H. nortieri*, viz. *H. nortieri* var. *globosiflora* (G. G. Smith) Bayer, (*Haw. Handb.* 119 (1976), Bayer). This differs not only in the globose flower, for which it is named, but also in the leaf form, being more inflated towards the tip of the leaf. This variety occurs with the characteristic globose flower only in the Botterkloof between Clanwilliam and Calvinia. Identical plants are found at the foot of the Van Rhynsdorp Pass, and on the Knersvlakte as far north-west as Komkans, but without the globose flower: this is where this variety meets the type.

H. nortieri var. *globosiflora*

H. notabilis von Poelln., *Fedde's Repert. Spec. Nov.* 45:134 (1938); *Journ S.A. Bot.* 14:56 (1948), G. G. Smith; *Cact. Amer.* 41:58 (1969), J. R. Brown; *Cact. Succ. Journ. NSW.* 8:9 (1971), Bayer; *Haw. Handb.* 141 (1976), Bayer

Bayer reduces this species to varietal rank under *H. magnifica.*

H. obtusa Haw. emend Uitew., *Succulenta* 49 (1948); *Phil. Mag.* 282 (1825), Haworth; *Cact. Amer.* 36:75 (1964), J. R. Brown; & 46:166 (1974), Bayer & Pilbeam; *Haw. Handb.* 141 (1976), Bayer

This species has been returned to its proper place as a variety at most of *H. cymbiformis*. Bayer has discounted all the former varieties of *H. cymbiformis* in his Handbook, so that this species is regarded as synonymous with the type, but some forms are maintained herein under that species, including this one at varietal level. Other varieties of *H. obtusa* resolve under *H. cooperi.*

H. nortieri var. *nortieri*

H. cv. ollasonii

H. cv. ollasonii G. Ren Hayes, *Cact. Succ. Journ. NSW.* 6:25 (1971)

This cultivar name has been given to a chance seedling found in the collection of Mr L. O. Ollason

H. nortieri var. *nortieri*

near Sydney in Australia; its origins are unknown, although its appearance suggests *H. cooperi* fa. *pilifera* crossed with *H. retusa* or something similar. It is a vigorous grower with shining, dark green leaves, translucent in the end-areas.

H. otzenii G. G. Smith, *Journ. S.A. Bot.* 11:72 (1945); *Haw. Handb.* 142 (1976), Bayer

There is confusion as to the identity of Smith's species, which, as pictured in the original description, seems to be near *H. turgida*, or *H. laetevirens*. Plants received from the type locality do not really match up to Smith's description, and accord with *H. mutica*. Either way the name warrants submerging, and is undoubtedly a source of confusion. Matters are not helped by the fact that no type was deposited as stated in Smith's description. Bayer rightly discards it in his Handbook.

There is a rapidly growing, heavily-clustering plant under this name invading collections within the last few years; its origin is unkown, but it is a misnomer, being probably nearest to *H. mirabilis* subsp. *mundula*, although the rapidity of growth suggests hybrid origins.

H. pallida Haw., *Revis. Pl. Succ.* 56 (1821); *The First Fifty Haw.* 26 (1970), Pilbeam; *Cact. Succ. Journ. NSW.* 8:9 (1971), Bayer; *Nat. Cact. Succ. Journ.* 27:10 (1972), Bayer; *The Second Fifty Haw.* 6 (1975), Pilbeam; *Haw. Handb.* 143 (1976), Bayer

Bayer discounts this species as probably of garden origin, probably with *H. herbacea* parentage, and *H. pallida* var. *paynei* as a small for mof *H. herbacea*.

H. papillosa (Salm-Dyck) Haw., *Suppl. Pl. Succ.* 58 (1819); *Pl. Succ. Hort. Dyck.* 12 (1816), Salm-Dyck; *Monogr.* 6:4 (1836–63), Salm-Dyck; *Cact. Amer.* 28:31 (1956), J. R. Brown; & 43:157 (1971), Bayer; *Cact. Succ. Journ. NSW.* 8:9 (1971), Bayer; *The First Fifty Haw.* 26 (1970), Pilbeam; *Haw. Handb.* 143 (1976), Bayer

Bayer submerged this species with *H. margaritifera*, discounting *H. papillosa* var. *semipapillosa* also as a variant of *H. margaritifera*. But C. L. Scott has revealed that *H. pumila* is the prior name for this species.

H. paradoxa von Poelln., *Fedde's Repert. Spec. Nov.* 33:240 (1933); *Des. Pl. Life* 9:90 (1937); & 43:104 (1938), von Poellnitz; *Cact. Amer.* 10:7 (1938), J. R. Brown; *Journ. S.A. Bot.* 14:55 (1948), G. G. Smith; *The First Fifty Haw.* 27 (1970), Pilbeam; *Haw. Handb.* 143 (1976), Bayer; *Nat. Cact. Succ. Journ.* 32:18 (1977), Bayer

Bayer submerges this species under *H. magnifica* at varietal level.

H. parksiana von Poelln., *Fedde's Repert. Spec. Nov.* 41:205 (1937); *Cact. Journ.* 5:34 (1936) & 6:19 (1937),

H. parksiana

H. parksiana

von Poellnitz; *Des. Pl. Life* 10:48 (1938), von Poellnitz; *Journ. S.A. Bot.* 55:14 (1948), G. G. Smith; *The Second Fifty Haw.* 43 (1975), Pilbeam; *Haw. Handb.* 143 (1976), Bayer

Section Loratae subsection Loratae

This miniature in the genus is a delight, assuming that you delight in the unusual, difficult and minuscule. Its tiny rosette, barely 2 or 3cm across, usually makes little height, reluctantly offsetting from the base to form large clusters after many years. The dark green, almost black leaves recurve slightly in the manner of the cactus *Ariocarpus kotschoubeyanus* and are encrusted with minute, concolorous tubercles. The flowers are borne on short stems of only 10cm or so, with only a few flowers to each stem (6 or 8). The species was named erroneously by von Poellnitz in honour of a non-existent collector by the name of Parks: in fact the collector's number ascribed to it by F. R. Long (Parks 636/32) referred to the Parks Department of Port Elizabeth of which he was superintendent. It comes from the Mossel Bay and Great Brak area, under shrubs hidden in fallen debris or among mosses and lichens. Some plants intermediate between this species and *H. floribunda* were recently found, which confirmed the misplacing of this species in the Section Retusae.

This species seems to grow better in partial shade from the brightest sun, but I have found it a difficult species to grow well for any length of time.

H. peacockii Bak., *Journ. Linn. Soc.* 18:202 (1880); *Pflanz.* 4.38:83 (1908). Berger; *Nat. Cact. Succ. Journ.* 28:80 (1973), Bayer; *Haw. Handb.* 144 (1976), Bayer

Bayer dismissed this species, unknown for decades, as a variant of *H. coarctata*.

H. pearsonii C. H. Wright, *Kew Bull.* 8:365 (1907); *Haw. Handb.* 144 (1976), Bayer; *Aloe* 18:7 (1980), Scott; *New. Haw. Handb.* 37 (1982), Bayer

Although Scott has recently suggested that this is the prior name for what is commonly known as *H. decipiens*, based on a painting in the Herbarium Library at Kew, it is not upheld here, as examination of the painting in question leaves considerable doubt as to its application, not least because of its poor execution; there are also indications in the painting of a regular line of bristles on the keels of the leaves, *not* a common feature of *H. decipiens*. It is regarded as insufficiently identifiable.

H. perplexa von Poelln., *Kakteenk.* 67 (1938); *Haw. Handb.* 145 (1976), Bayer

F. R. Long wrote to von Poellnitz at the time this species was named to say that it was a natural hybrid. Bayer has elicited in correspondence that it is such, between *H. cymbiformis* and *H. angustifolia*.

H. picta von Poelln., *Fedde's Repert. Spec. Nov.* 44:133 (1938); *Des. Pl. Life* 10:126 (1938), von Poellnitz; *Cact. Amer.* 12:174 (1940), J. R. Brown; *Aloe* 11:8 (1973), C. L. Scott; & 12:89 (1974), Bayer; *The Second Fifty Haw.* 44 (1975), Pilbeam; *Haw. Handb.* 145 (1976), Bayer

Bayer has placed this species into synonymy with *H. emelyae*, which is a pity, since the name 'picta', meaning painted, well conveyed the coloration of which this species is capable if grown in good light conditions.

H. pilifera Bak., *Saund. Refug. Bot.* 4:t.234 (1870); *Journ. Linn. Soc.* 18:214 (1880); *Fl. Cap.* 6:353 (1896), Dyer; *Succulenta* 101 (1937), Uitew.; *Cact. Amer.* 46:166 (1974), Bayer & Pilbeam; *Haw. Handb.* 145 (1976), Bayer

This well-known species was restored to its original specific status after an unhappy interlude of 26 years as a variety of *H. obtusa*. The reinstatement was made by Bayer and the author in a paper in the U.S. Society's journal (ref. above). In this paper it was maintained that *H. pilifera* was a widespread, variable species, which had given rise to several varieties erected without 'real field study of the variation of the population as a whole'. After extensive study in the field Bayer concluded that *H. pilifera* var. *dielsiana*, *H. pilifera* var. *columnaris* and *H. pilifera* var. *salina* fell into synonymy with the type, and that for the moment *H. pilifera* var. *gordoniana* and *H. pilifera* var. *stayneri* should maintain their former specific status. But subsequently Bayer took the bull by the horns in his Handbook and reduced the whole lot under *H. cooperi*, the prior name in this complex (this was anticipated in a footnote in the above-mentioned paper in the U.S. journal). See under *H. cooperi*.

H. planifolia Haw., *Phil. Mag.* 282 (1825); *Monogr.* 11:2 (1836–63), Salm-Dyck; *Fedde's Repert. Spec. Nov.* 45:161 *et seq.* (1938), von Poellnitz; *Cact. Amer.* 10:147 (1939), J. R. Brown; *The First Fifty Haw.* 27 (1970), Pilbeam; *Haw. Handb.* 145 (1976), Bayer

This species was reduced by Bayer to synonymy with *H. cymbiformis*. It is upheld herein at form level under that species. The only one of the numerous varieties erected by von Poellnitz to be recognized is *H. planifolia* var. *transiens*, which becomes a variety of *H. cymbiformis*. The rest merely represent variants of a very variable species.

H. poellnitziana Uitew., *Cact. Vetpl.* 5:137 (1939); *Des. Pl. Life* 18:116 (1946), J. R. Brown; *Cact. Amer.* 43:157 (1971), Bayer; *Haw. Handb.* 146 (1976), Bayer Section Margaritiferae

This well-marked species, with affinities to *H. minima*, has tentatively been maintained by Bayer in his Handbook. He comments that the nearest population of *H. minima* is 30km away, and that it is ecologically unique in situation at the western extremity of the range for *H. pumila*, with which it is also related. It resembles an attenuated *H. minima*, thickly tubercled on both sides of the leaves. It is reported from Drew, west of Swellendam. It is little known in cultivation as yet.

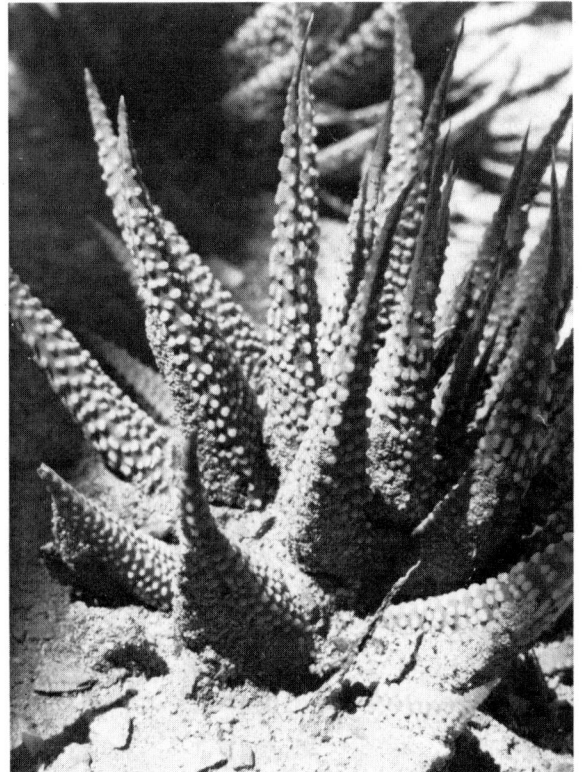

H. poellnitziana

H. pseudogranulata von Poelln., *Fedde's Repert. Spec. Nov.* 41:208 (1937); *Haw. Handb.* 147 (1976), Bayer

Bayer refers this species to probable synonymy with *H. venosa* subsp. *tessellata*, in one of its many and varied forms.

H. pubescens Bayer, *Journ. S.A. Bot.* 38:125 (1972); *Aloe* 11:8 (1973) C. L. Scott; & 12:89 (1974), Bayer; *The Second Fifty Haw.* 12 (1975), Pilbeam; *Haw. Handb.* 147 (1976)
Section Retusae subsection Turgidae

This is a recently described species, which comes from the Sandberg hills, 12km south-south-east of Worcester, having connections with *H. magnifica* var. *maraisii* and the *H. herbacea* complex. It is a beauty for the Haworthia connoisseur, being small (2.5 to 4cm diameter) and heavily covered all over the leaves' surfaces with small, translucent tubercles, surmounted by small translucent bristles. The dark green, triangular leaves colour to a grey-brown-purple in full light. For such a small rosette there is a surprising number of leaves, about 40 to 50 when full-grown, densely packed into a short, thick stem half the diameter of the whole rosette. It tends to stay low on the surface of the soil, making only 1 or 2cm in height, and seldom offsetting, growing extremely slowly. Its habit and habitat suggest care with watering, especially overhead, as water in such a pubescent, packed rosette would tend to linger and might cause rotting of the very thick stem below.

H. pubescens

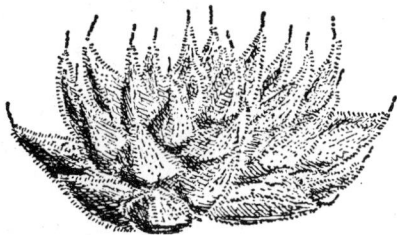

H. pubescens

H. pulchella Bayer, *Journ. S.A. Bot* 39:232 (1973); *The Second Fifty Haw.* 12 (1975), Pilbeam; *Haw. Handb.* 147 (1976), Bayer
Section Loratae subsection Loratae

Found barely ten years ago and described in 1973, this species is another small gem, coming from an area extending from Avondrust to Nongaspoort, south-east of the Touws River, growing close to *H. arachnoidea*, but quite separate from possible allies in the forms of *H. angustifolia*, *H. herbacea* or some forms of *H. magnifica*. It forms small, single rosettes, seldom offsetting, up to 4.5cm across, with up to 45 leaves to the rosette; the stem is narrow, only 5mm wide. The leaves are constantly incurved and more or less evenly lanceolate with a strong keel (sometimes there are two keels) armed with reflexed teeth, as are the margins in the upper parts of the leaves. These prominent teeth give the species its attraction, especially when the plant is dry and the leaves strongly incurving. It is not quick-growing, but seems not difficult to cultivate, although it should be borne in mind that the species comes from a very dry area, so that prolonged overwatering will be detrimental.

H. pulchella

H. pulchella

H. pumila (L.) Duv., *Pl. Succ. Hort. Alenc.* 7 (1809); *Sp. Pl.* 322 (1753), Linnaeus; *Hort. Med. Amst.* 2:19.t.10 (1701), Commelin; *Monogr.* 6:5 (1836–63), Salm-Dyck—*Aloe margaritifera*; *Cact. Amer.* 28:51

(1956), J. R. Brown—*H. margaritifera*; *Haw. Handb.* 133 (1976), Bayer—*H. margaritifera*; *Aloe* 16:44 (1978), C. L. Scott.
Section Margaritiferae

As explained under the entry for *H. margaritifera*, this is the prior name for the species which has been known for the last nearly 200 years as *H. margaritifera*. Sad as it may seem to lose a popular and descriptive name, the rules clearly expounded by Colonel Scott leave us no alternative but to accept. *H. pumila* is anomalously so named since it is probably the biggest species of Haworthia, because it was originally described as an Aloe (*pumila* means small or dwarf).

Also formerly reduced to synonymy with this species are *H. papillosa* and its variety *semipapillosa*.

If you are the sort of collector who likes the big species in any genus then this is the one for you, making rosettes nearly the size of a large melon in time. Its large triangular leaves are thick and fleshy and covered, though not so densely as in some forms of *H. minima*, with ice-drop, white tubercles which stand out from the surface of the leaf prominently; the upper surface of the leaf is sometimes not tuberculate or has much smaller tubercles, sparsely arranged compared with the backs of the leaves. The flower stem is characteristic of the subgenus to which it is allocated, thick and branching and carrying some hundreds of flowers on a large specimen. It comes from a widespread area in the Robertson area of the Karoo and the western part of the Little Karoo, sometimes hybridizing with neighbouring populations of *H. marginata* and producing intermediate plants, and with *H. minima*, with similar results.

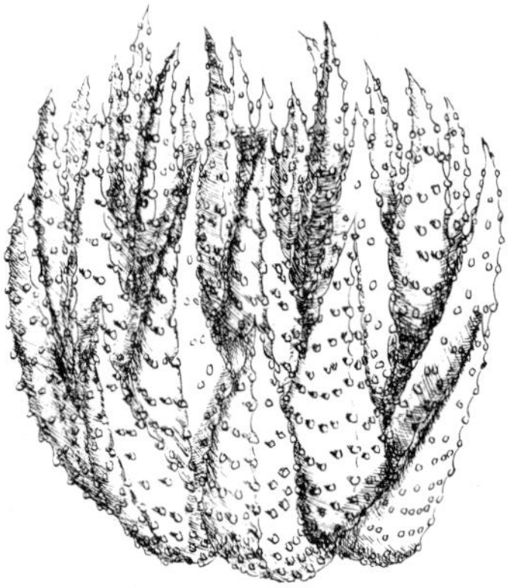

H. pygmaea von Poelln., *Fedde's Repert. Spec. Nov.* 27:132 (1930); & 41:208 (1937), von Poellnitz; *Kakteenk.* 173 (1937), von Poellnitz; *Cact. Amer.* 13:75 (1941), J. R. Brown; & 30:133 (1958), J. R. Brown; *Excelsa* 5:83 (1975), Bayer
Section Retusae subsection Retusae

H. pumila

H. pygmaea fa. *pygmaea*

H. pumila

H. pygmaea fa. *major*

There has been some doubt as to the true identity of this species, for a long time little known in collections, except often misnamed. It was reported from Great Brak, and Bayer has affirmed its occurrence only at Mossel Bay, Great Brak, although von Poellnitz's reference to its occurrence at De Rust indicates perhaps its affinities to scabrous forms of *H. emelyae* in the Oudtshoorn area. It is, as the name indicates, one of the smallest of the section with strongly recurved leaves, more or less papillate (see form below) and colouring well to reddish hues in full light. It is slow-growing, but will offset from the base in time. Watering should be moderate, and a fairly exposed situation will show the species at its best. The very papillose form, usually seen in collections labelled *H. asperula*, is referable here, and is maintained as a form of the species.

H. pygmaea fa. *major*

H. pygmaea fa. *crystallina* fa. nov. *A typo differt parte folii apicali truncata valde papillosa.* Differs from the type in having leaves with very papillate end-areas. Type deposited at Kew.

H. pygmaea fa. *crystallina*

Most commonly seen in cultivation recently is a larger form than the type, also from Great Brak, with grey-green leaves, the rosette getting to twice the size of the type, to about 6 cm in diameter. It is described here as a new form: *H. pygmaea* fa. *major*, fa. *nov.*, *a typo differt folius majoribus.* Differs from the type in having larger leaves.

H. radula (Jacq.) Haw., *Syn. Pl. Succ.* 93 (1812); *Hort. Schoenb.* 4:11 (1804), Jacquin; *Monogr.* 6:8 (1836–63), Salm-Dyck; *Des. Pl. Life* 7:33 (1935), J. R. Brown; *The First Fifty Haw.* 30 (1970), Pilbeam; *Haw. Handb.* 148 (1976), Bayer
Section Coarctatae subsection Attenuatae

H. radula

H. radula

This is a fine species with long, attenuated leaves, covered with tiny tubercles, giving the leaf a rough texture as the name implies. It is common in collections, but not commonly well-grown to bring out the best coloration and growth in the species. Given favourable conditions, sufficient water and a fairly exposed position with good drainage, it will not tend to dry up at the leaf-tips as is the wont of this sort of long-leaved Haworthia, and will colour to reddish-brownish-purple, throwing into relief the whiteness of the tiny tubercles, especially at the base of the leaves, where they tend to be a little larger and more inclined to white coloration. It will grow into large tangled clumps of intertwining leaves, curving this way and that. Some forms of *H. attenuata* come very close to this species in leaf-character, and Bayer reports that they both grow together in the Hankey area of the Gamtoos Valley, but apparently in separate small populations rather than randomly associated. *H. radula* always has minute crowded tubercles, and there do not seem to be intermediate stages between this and the large, more spaced tubercles of *H. attenuata*.

H. ramosa G. G. Smith, *Journ. S.A. Bot.* 10:22 (1944); *Cact. Amer.* 25:32 (1953), J. R. Brown; *The Second Fifty Haw.* 46 (1975), Pilbeam; *Haw. Handb.* 149 (1976), Bayer

This species has been reduced by Bayer to synonymy with *H. cymbiformis*. It is retained at form level herein.

H. recurva Haw., *Syn. Pl. Succ.* 94 (1812); *Fl. Cap.* 6:345 (1896), Dyer; *Haw. Handb.* 149 (1976), Bayer

This name has hitherto been regarded as applying to a thick-leaved species in the style of *H. tessellata*, but with opaque upper surfaces to the leaves, and making

short columns of strongly recurving leaves, the back surface rough from concolorous tubercles. Bayer in his Handbook maintained that this and *H. tessellata* were synonymous, and reduced both under the former named *H. recurva* to subspecific status under *H. venosa*, the oldest name of the three. But he has relented under pressure and reinstated the popular and apt name *H. tessellata* in preference to *H. recurva*, but still as a subspecies of *H. venosa*—see under the latter name.

H. reinwardtii (Salm-Dyck) Haw., *Revis. Pl. Succ.* 53 (1821); *Obs. Bot.* 37 (1820), Salm-Dyck; *Monogr.* 6:16 (1836–63), Salm-Dyck; *Cact. Amer.* 12:43 (1940), J. R. Brown
Section Coarctatae subsection Coarctatae

This species, with its very wide variation of leaf patterning and size of stem, has given more ammunition to the scoffers at the over-splitting of this genus

H. reinwardtii var. *reinwardtii* fa. *reinwardtii*

than any other, except perhaps *H. attenuata*. But *H. reinwardtii* varieties in the main were erected after considerable study in the field, especially those described by G. G. Smith, and were based mostly on populations rather than individuals, although followers tend to stick slavishly to the type described rather than allow for variation in these populations. It must be admitted, however, that there are more names than are needed for *H. reinwardtii*, and Bayer's rationalization of this and the species *H. coarctata* in 1973 brought some long overdue sense to the muddle of over 20 varietal names ascribed to *H. reinwardtii*. In this paper Bayer maintained that some varieties of *H. reinwardtii* had more affinity with *H. coarctata*; i.e. *H. reinwardtii* var. *adelaidensis*, *H. reinwardtii* var. *tenuis*, *H. reinwardtii* var. *bellula* and *H. reinwardtii* var. *riebeekensis*. Bayer's opinion was that the other varieties equated with *H. reinwardtii* and that none warranted upholding at varietal level. But he conceded that some were 'striking collector's pieces', and could be retained at form level. In *The Second Fifty Haworthias* the author upheld at form level the former varieties *H. reinwardtii* var. *brevicula*, *H. reinwardtii* var. *diminuta*, *H. reinwardtii* var. *chalumnensis*, *H. reinwardtii* var. *kaffirdriftensis* and *H. reinwardtii* var. *zebrina*. Bayer in his Handbook retained *H. reinwardtii* var. *brevicula* (as a variety, note), and took it to embrace *H. reinwardtii* var. (or fa.) *diminuta*; he also upheld the other forms mentioned immediately above, and added *H. reinwardtii* fa. *olivacea* (former *H. reinwardtii* var. *olivacea*).

This leaves collectors somewhat confused, and I doubt that many labels were altered, in spite of Bayer's thorough treatment of the species. The following is an attempt to rationalize for collectors the current state of the species, as seen by the author, and to point out which forms are worth seeking out.

H. reinwardtii var. *reinwardtii* fa. *reinwardtii*

H. reinwardtii var. *reinwardtii* fa. *kaffirdriftensis*

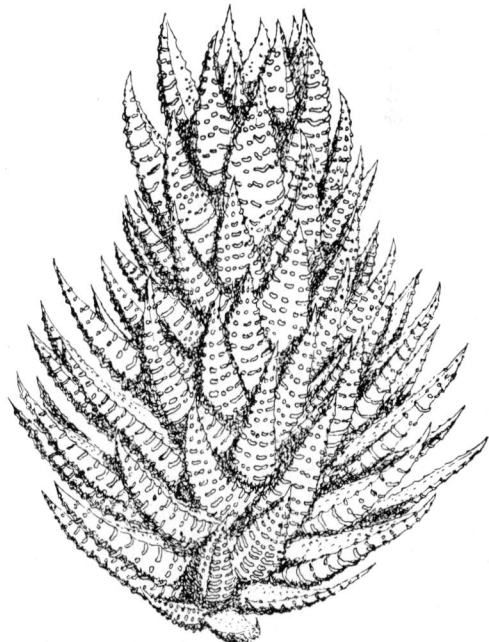

H. reinwardtii var. *reinwardtii* fa. *chalumnensis*

H. reinwardtii var. *reinwardtii* fa. *kaffirdriftensis*

Firstly the type, *H. reinwardtii* var. *reinwardtii* fa. *reinwardtii*, to give it its full title, for which one can do no better than to refer the reader to Salm-Dyck's Monograph illustration. The type in present-day collections is indeterminate, but is perhaps best typified by former varieties close to the original concept and not maintained by Bayer as of sufficient distinction to retain as separate entities, such as G. G. Smith's *H. reinwardtii* var. *peddiensis* or *H. reinwardtii* var. *grandicula* (*Journ. S.A. Bot.* 9:94:1943 & 10:12 (1944)).

H. reinwardtii var. *reinwardtii* fa. *olivacea*

H. reinwardtii var. *reinwardtt* fa. *zebrina*

Secondly let us dispose of Baker's two varieties *H. reinwardtii* var. *major* and *H. reinwardtii* var. *minor* (*Journ. Linn. Soc.* 18:202 (1880)), whose descriptions are woefully inadequate in view of the subsequently revealed large variation of the species; they were merely described as larger and smaller than the type. Bayer did not uphold these two old varietal names in his Handbook, pointing out that the inadequate description also made it impossible to determine whether they were proper to *H. reinwardtii* or *H. coarctata*.

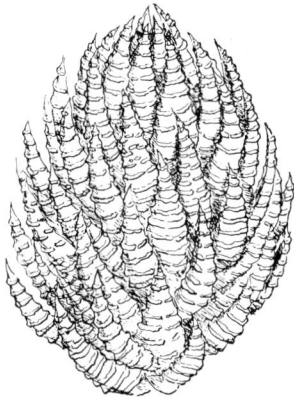

H. reinwardtii var. *reinwardtii* fa. *zebrina*

H. reinwardtii var. *adelaidensis* von Poelln. (*Beitr. Sukk. u. Pfl.* 43 (1940)) is now placed under *H. coarctata*.

H. reinwardtii var. *archibaldiae* von Poelln. (*Fedde's Repert. Spec. Nov.* 41:210 (1937)) was discounted at any level by Bayer, being insufficiently discrete in the wild to consider as a separate entity. It is perhaps closest in appearance to *H. reinwardtii* var. *chalumnensis*.

H. reinwardtii var. *bellula* G. G. Smith (*Journ. S.A. Bot.* 11:70 (1945)), rejected by Bayer, is maintained herein as a form of *H. coarctata* subsp. *adelaidensis*.

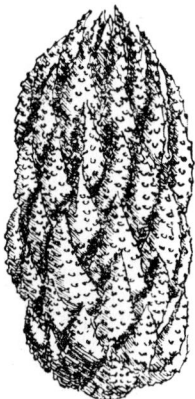

H. reinwardtii var. *brevicula*

H. reinwardtii var. *brevicula* G. G. Smith (*Journ. S.A. Bot.* 10:11 (1944)) is the most prominently marked of the smaller end of the *H. reinwardtii* complex, with bright white, solitary tubercles on dark green leaves. The stems grow to about 8cm tall and are about 2 to 3cm wide. Bayer maintains this variety, taking it also to cover the later described *H. reinwardtii* var. *diminuta*, as both were reported from Frasers Camp, east of Grahamstown, and are very close in appearance. In common with other small variants of this species it seems to present more difficulty in cultivation, being more slow-growing than the larger forms and making a weaker root system. A gritty compost and a lighter hand on the watering-can is indicated for successful cultivation.

H. reinwardtii var. *brevicula*

H. reinwardtii var. *chalumnensis* is for me the most handsome of the larger-growing forms of this species, making massive clumps in the wild on the banks of the Chalumna River, 30 miles west of East London. It is strong-growing and colours purple-red in full light showing off well the prominent, white, nearly confluent bands of tubercles on the backs of the leaves. The individual stems will grow to 13cm or more tall and are 6 to 7cm wide. The leaves tend to stand out from the stem more than most other forms of the species. Bayer discards this variety in his rationalization of the species in 1973, but it was subsequently upheld in *The Second Fifty Haworthias* by the author, at form level, and by Bayer in his Handbook. It presents little difficulty in cultivation except perhaps in keeping up with its rapid growth by allowing it room to develop into a large clump, when, with its beautiful colouring in full light it makes a very impressive plant.

H. reinwardtii var. *chalwinii* (Marl & Berg.) Res., (*Notizbl. Bot. Gart. Mus. Berlin.* 4:247 (1906); *Mem. Soc. Brot.* 67 (1943), Resende; *Cact. Amer.* 12:59 (1940), J. R. Brown; *The Second Fifty Haw.* 26 (1975) (fig. on left p. 27), Pilbeam; *Haw. Handb.* 106 (1976), Bayer). This well-known taxon is maintained under *H. coarctata* at form level.

H. reinwardtii var. *reinwardtii* fa. *chalumnensis*

H. reinwardtii var. *committeesensis* G. G. Smith (*Journ. S.A. Bot.* 9:93 (1943); *Cact. Amer.* 40:142 (1968), J. R. Brown; *The First Fifty Haw.* 31 (1970), Pilbeam; *Haw. Handb.* 107 (1976), Bayer). Referred to synonymy by Bayer with *H. coarctata*, as it did not represent a marked discontinuity with the species.

H. reinwardtii var. *conspicua* von Poelln. (*Fedde's Repert. Spec. Nov.* 41:210 (1937); *Cact. Amer.* 18:157 (1946), J. R. Brown; *The Second Fifty Haw.* 26 (1975) (fig. on right p. 27), Pilbeam; *Haw Handb.* 109 (1976), Bayer). Maintained at form level under *H. coarctata*.

H reinwardtii var. *diminuta* G. G. Smith, (*Journ. S.A. Bot.* 14:52 (1949); *Cact. Amer.* 31:151 (1959), J. R. Brown; *The Second Fifty Haw.* 47 (1975), Pilbeam; *Haw. Handb.* 113 (1976), Bayer). Referred to synonymy with *H. reinwardtii* var. *brevicula*.

H. reinwardtii var. *fallax* von Poelln. (*Fedde's Repert. Spec. Nov.* 31:83 (1932); & 41:209 (1937); *Cact. Amer.* 32:189 (1960), J. R. Brown; *Haw Handb.* 116 (1976), Bayer). This variety is referred by Bayer to synonymy with *H. coarctata*.

H. reinwardtii var. *grandicula* G. G. Smith, (*Journ. S.A. Bot.* 10:12 (1944); *Haw Handb.* 120 (1976), Bayer). This variety was described by Smith from plants collected east of Kaffir Drift in the Peddie Division. It has stems up to about 13cm long and about 4 or 5cm broad. It is not upheld at any level and is referred to synonymy with *H. reinwardtii*.

H. reinwardtii var. *huntsdriftensis* G. G. Smith, (*Journ. S.A. Bot.* 10:14 (1944); *Cact. Amer.* 33:118 (1961), J. R. Brown; *Haw. Handb.* 122 (1976), Bayer). This variety comes from 'several miles south west of Hunts Drift on the Fish River', and is not remarkably distinct in Bayer's opinion from the type. The tubercles are small and do not show up as much as in other more distinctive varieties. There is also some doubt about the plants described and their locality. The name is not upheld at any level.

H. reinwardtii var. *kaffirdriftensis* G. G. Smith (*Journ. S.A. Bot.* 9:96 (1948); *Cact. Amer.* 31:60 (1959), J. R. Brown; *The Second Fifty Haw.* 48 (1975), Pilbeam; *Haw. Handb.* 126 (1976), Bayer). This variety shines through as a deserving collector's piece, and has been described as the most attractive of the species. This is due to the clear white tubercles, which tend to coalesce in longitudinal rows on the backs of the leaves, forming more or less continuous vertical ridges, more prominent and thicker at and nearer the keel. It makes modest sized stems, about 12cm tall and about 3.5cm in diameter, and it clusters from or near to the base of each stem. In full light the ground-colour of the leaves becoms dark green to purple-red, and they stay incurving and clasping the stem. It was upheld at form level in *The Second Fifty Haworthias* by the author, and Bayer ratifies this in his Handbook. It occurs as the name indicates near Kaffir Drift in the Peddie Division.

H. reinwardtii var. *olivacea* G. G. Smith (*Journ. S.A. Bot.* 10:142 (1944); *Cact. Amer.* 34:181 (1962), J. R. Brown; *Haw. Handb.* 142 (1976), Bayer). Bayer maintains this variety at form level in his Handbook. It has the distinction of unusual-coloured stems—olive-green, the colour extending into the tubercles, which are greenish-white; in full light conditions it will become intensely orange-brown. It comes from Kaffir Drift, and present no difficulty in cultivation, making clumps fairly quickly, offsetting at and near the base, with stems about 8cm tall and 4cm wide, and leaves rather narrower than other forms of this size.

H. reinwardtii var. *peddiensis* G. G. Smith (*Journ. S.A. Bot.* 9:94 (1943); *Cact. Amer.* 24:165 (1952), J. R. Brown; *Haw. Handb.* 144 (1976), Bayer). This is another little-distinguished variety from the Peddie Division, which represents as typical a form as any. Bayer discards it as indistinct and with the comment that it seems to have been collected virtually from within the town of Peddie, and does not seem to exist there any more. It is not upheld at any level.

H. reinwardtii var. *pulchra* von Poelln. (*Fedde's Repert. Spec. Nov.* 41:209 (1937); *Haw. Handb.* 148 (1976), Bayer). This variety was recorded from Kaffir Drift and has nothing to distinguish it from others very similar in appearance. It is not upheld at any level.

H. reinwardtii var. *riebeekensis* G. G. Smith, (*Journ. S.A. Bot.* 10:16 (1944); *Cact. Amer.* 34:14 (1962), J. R. Brown; *The First Fifty Haw.* 32 (1970), Pilbeam; *Haw. Handb.* 150 (1976), Bayer). This variety has been referred to synonymy with *H. coarctata* var. *tenuis*.

H. reinwardtii var. *tenuis* G. G. Smith (*Journ. S.A. Bot.* 14:51 (1948); *Cact. Amer.* 31:112 (1959), J. R. Brown; *Haw. Handb.* 161 (1976), Bayer). This variety has been referred to *H. coarctata* as a variety.

H. reinwardtii var. *triebneri* Res. (*Mem. Soc. Brot.* 80 (1943); *Cact. Amer.* 40:143 (1968), J. R. Brown; *Haw. Handb.* 177 (1976), Bayer). It is difficult to determine what Resende was describing, and its origins. Plants seen under this name have been hybrids between apparently *H. reinwardtii* and *H. attenuata*, intermediate in character.

H. reinwardtii var. *valida* G. G. Smith (*Journ. S.A. Bot.* 9:98 (1943); *Cact. Amer* 34:83 (1962), J R.

Brown; *Haw. Handb.* 165 (1976), Bayer). Collected 10 miles south-west of Peddie, this variety was distinguished by its comparatively large tubercles placed singly and prominently on the backs of the leaves. It is not upheld at any level by Bayer, and is not sufficiently distinguished to warrant retention.

H. reinwardtii var. *zebrina* G. G. Smith (*Journ. S.A. Bot* 10:18 (1944); *Cact. Amer.* 24:131 (1952), J. R. Brown; *The Second Fifty Haw.* 48 (1975), Pilbeam; *Haw. Handb.* 169 (1976), Bayer). About this variety there is no argument whatever, since it is one of the most distinct and distinguished in the species, with bands of coalescing tubercles on the backs of the leaves horizontally, reminiscent of *H. fasciata*. Although Smith chose the more continuously banded form when describing this variety, the continuous nature of the banding is variable in the wild, and plants with quite separate tubercles occur in the same colony. This makes its reduction to form level in *The Second Fifty Haworthias* by the author, ratified by Bayer in his Handbook, a logical placing for this attractive, sought-after variation. It was collected originally by Smith on the banks of the Fish River, south-east of Kaffir Drift. Its only difficulty in cultivation is its desirability, which tends to mean that offsets are removed as fast as they appear for other eager collectors; it would, if allowed to, make a small clump fairly quickly.

In summary, the species resolves as follows:
H. reinwardtii var. *reinwardtii*.
H. reinwardtii var. *reinwardtii*. fa. *chalumnensis*.
H. reinwardtii var. *reinwardtii*. fa. *kaffirdriftensis*.
H. reinwardtii var. *reinwardtii*. fa. *olivacea*.
H. reinwardtii var. *reinwardtii*. fa. *zebrina*.
H. reinwardtii var. *brevicula*.

H. resendeana von Poelln., *Des. Pl. Life* 10:225 (1938); *Cact. Amer.* 14:165 (1942), J. R. Brown; *The First Fifty Haw.* 32 (1970), Pilbeam

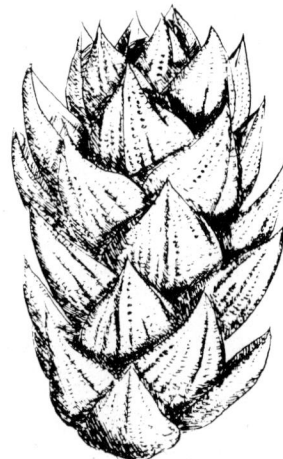

H. resendeana

There are two plants in cultivation vying for this name, and it is not possible with certainty to determine which is that originally described by von Poellnitz. Neither is particularly distinguished, and both are almost certainly of hybrid origin. In fact Bayer dismisses the name in his Handbook as an 'unknown triploid hybrid'. It is however known in collections, and for the record the most usual contenders are illustrated, but its occurrence in the wild is doubted, and it should be quietly laid to rest.

H. reticulata Haw., *Trans. Linn. Soc.* 7:9 (1804)— *Aloe reticulata*; *Syn Pl. Succ.* 94 (1812); *Monogr.* 10:1 (1836–63), Salm-Dyck; *Cact. Amer.* 18:54 (1946), J. R. Brown; *Nat. Cact. Succ. Journ.* 27:10 (1972), Bayer; *The Second Fifty Haw.* 5–7 (1975), Pilbeam; *Haw. Handb.* 150 (1976), Bayer; *New Haw. Handb.* 52 (1982), Bayer

Section Retusae subsection Turgidae

This species has become the repository for several erected in the 1930s by von Poellnitz. Bayer, after extensive field studies of this and the neighbouring species, *H. herbacea*, upheld only these two at specific level. Former species and varieties declared synonymous with *H. reticulata* were: *H. reticulata* var. *acuminata*, *H. hurlingii* (subsequently in his 1982 Handbook upheld at varietal level), *H. hurlingii* var. *ambigua*, *H. haageana*, *H. haageana* var. *subreticulata*, *H. intermedia* and *H. subregularis*.

The species resolves as follows, with *H. subregularis* retained at varietal level to represent the larger forms of the species reported from the Lower Hex River Valley.

H. resendeana

H. reticulata var. *reticulata*

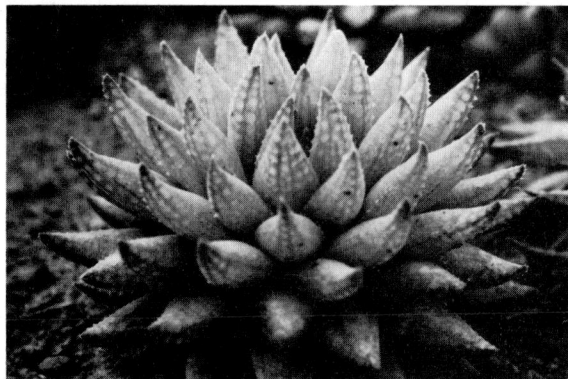

H. resendeana

H. reticulata var. *reticulata* has rosettes about 4cm in diameter, clumps readily, and colours well to red tones in full light.

H. reticulata var. *reticulata*

H. reticulata var. *hurlingii*

H. reticulata var. *subregularis*

H. reticulata var. *hurlingii* (von Poelln.) Bayer, is found between Robertson and Bonnievale. Rosettes of this variety are no more than 2 or 3cm in diameter, the leaves almost round in outline and when turgid like tiny netted berries. It will colour to intense shades of red in full light.

H. reticulata var. *subregularis* stat. nov. (syn. *H. subregularis* Bak., *Saund. Refug.* 4:t.232 (1870); & *H. haageana* von Poelln., *Fedde's Repert. Spec. Nov.* 28:104 (1930)). This variety has rosettes about 5cm in diameter or larger in cultivation, and forms large clumps in time, but more slowly than the other varieties. It too will colour to orange-red in full light. The species occurs over a wide area between Worcester and Robertson, north of the Breede River.

H. retusa (L.) Duv., *Pl. Succ. Hort. Alenc.* 7 (1809); *Sp. Pl.* 322 (1753), Linnaeus; *Syn. Pl.Succ.* 95 (1812), Haworth; *Aloe* 11:35 (1973), C. L. Scott; *The Second Fifty Haw.* 48 (1975), Pilbeam; *Haw. Handb.* 150 (1976), Bayer; *Excelsa* 8:46 (1978), Bayer
Section Retusae subsection Retusae

There is little doubt as to the concept of this species in the late eighteenth and nineteenth centuries: illustrations under this name at that time depict a large and thick-leaved plant, matching well with plants under this name in collections today. But perhaps because this is the type species of the Section Retusae, and the section name is easily transposed for the specific, there are a good many plants in collections under this name, which, although they belong to this section do not fall under the species. The confusion is not helped by the comparatively unknown varieties erected by G. G. Smith in the 1940s, which are difficult to identify with certainty, and do not seem to have got into collections at that time or since. They are *H. retusa* var. *solitaria* (*Journ. S.A. Bot.* 12:5 (1946)), *H. retusa* var. *densiflora* (*Journ. S.A. Bot.* 12:7 (1946)), and *H. retusa* var. *multilineata* (*Journ. S.A. Bot.* 12:3 (1946)), the last named being somewhat known, although plants seen in collections often do not match up to Smith's excellent photograph in his original description, being generally nearer the later described *H. geraldii*, which Bayer also places into synonymy with *H. retusa*. There is too *H. retusa* var. *mutica* (Haw.) Bak. (*Fl. Cap.* 6:346 (1896); *Rev. Pl. Succ.* 55 (1821), Haworth), but Bayer has restored this to its former status as a species; see under *H. mutica*.

H. retusa var. *retusa* fa. *retusa*

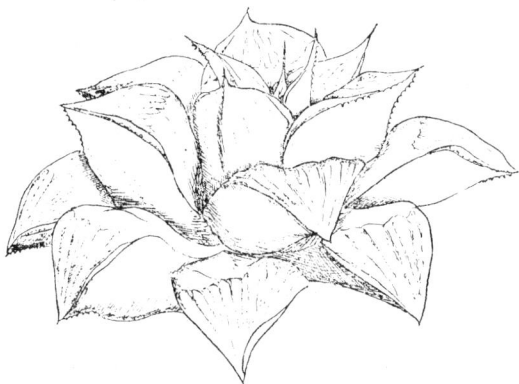

H. retusa var. *retusa* fa. *retusa*

Bayer in his Handbook says of *H. retusa* var. *solitaria* that it was subsequently found to be proliferous, and not as the name implied solitary (this in private correspondence with Mr. J. Dekenah), and that it is a variant of *H. retusa* found to the north-west of Riversdale on a high plateau, densely grassed, where these plants are scabrous, i.e. with a roughened surface to the leaves.

Of *H. retusa* var. *densiflora* Bayer says that the exact locality is not recorded and only given as Riversdale; he does not uphold the variety.

H. retusa var. *multilineata* is a variant with many lines in the end-areas of the leaves, and is a vigorous grower making large clumps of light-green-coloured rosettes, becoming red-brown in full light. It looks best if allowed to grow strongly, with room given for its vigorous roots, and sufficient breathing space for the many rosettes produced around the main rosette. Good light will prevent the rosettes from opening up too much, which this strong grower tends to do if given insufficient light. It comes from an area two miles north of Riversdale. It is maintained below as a form of the species.

Bayer in his 1976 Handbook recognized under *H. retusa* the former variety of *H. dekenahii* (viz. *H. dekenahii* var. *argenteo-maculosa*) but rejected the type, *H. dekenahii*, as a form of *H. retusa* inextricably involved with *H. turgida*. However in his revised Handbook 53 (1982) he resurrects *H. dekenahii* as the name correctly applied to the variety of *H. retusa*, i.e. *H. retusa* var. *dekenahii* (syn. *H. dekenahii*, *H. dekenahii* var. *argenteo-maculosa* and *H. retusa* var. *argenteo-maculosa*).

H. retusa var. *retusa* fa. *fouchei*

H. retusa var. *retusa* fa. *geraldii*

The other variety (at form level in the 1976 Handbook) which Bayer recognizes is *H. retusa* var. *acuminata*, distinguished by its very pointed leaves and slight roughening of the end-areas. This variety has been seen in Britain labelled variously as *H. paradoxa*, *H. retusa* var. *densiflora* and *H. magnifica*.

Further species relegated to synonymy in Bayer's Handbook include *H. asperula*, which has been a considerable source of confusion for years, associated by Scott with *H. pygmaea*, *H. paradoxa*, *H. magnifica* var. *maraisii*, *H. schuldtiana* and *H. pubescens*, and by von Poellnitz with plants from Great Brak (*H. pygmaea*), Bonnievale (*H. mutica*), Zebra (*H. emelyae*), Uniondale (*H. comptoniana*) and Barrydale (*H. magnifica* var. *maraisii*). Salm-Dyck's illustration in his Monograph (*Aloe asperula*) shows a plant which lends itself to interpretation as *H. pygmaea* more than anything else. With such doubt about the true identity of Haworth's plant it is best discarded as a source of confusion.

H. retusa var. *retusa* fa. *longebracteata*

H. retusa var. *retusa* fa. *multilineata*

Another species associated here is *H. fouchei*, which Bayer dismisses as an ecotypic variant of *H. retusa* occurring north-east of Riversdale, leading on to another variant called formerly *H. longebracteata*, on the coast. Both have more erect and narrower leaves than in the typical *H. retusa*, and cluster much more readily. *H. fouchei* is noted for its translucent end-areas to the leaves. *H. longebracteata* makes larger rosettes with distinct, white lines and flecking in the end-areas. Both are maintained as forms.

H. retusa var. *acuminata*

H. geraldii is also dismissed by Bayer as a variant of *H. retusa* again from north-east of Riversdale occurring near *H. fouchei*, and forming huge clumps. It is a distinctive form and is maintained as such below.

The species resolves as follows:

H. retusa (L.) Duv. var. *retusa* fa. *retusa*.

H. retusa var. *retusa* fa. *fouchei* (von Poelln.) Pilbeam, stat. nov. (syn. *H. fouchei* von Poelln., *Succulenta* 22:28 (1940)).

H. retusa var. *retusa* fa. *geraldii* (G. G. Smith) Pilbeam, stat. nov. (syn. *H. geraldii* Scott).

H. retusa var. *retusa* fa. *longebracteata* (G. G. Smith) Pilbeam, stat. nov. (syn. *H. longebracteata* G. G. Smith, *Journ. S.A. Bot.* 11:75 (1945)).

H. retusa var. *retusa* fa. *multilineata* (G. G. Smith) Pilbeam, stat. nov. (syn. *H. retusa* var. *multilineata* G. G. Smith, *Journ. S.A. Bot.* 12:3 (1946)).

H. retusa var. *acuminata* Bayer.

H. retusa var. *dekenahii* (G. G. Smith) Bayer (syn. *H. dekenahii* G. G. Smith).

H. retusa var. *dekenahii*

H. revendettii Uitew., *Cact. Vetpl.* 44 (1940); *Mem. Soc. Brot.* 92 (1943), Resende; *Cact. Amer.* 28:190 (1956), J. R. Brown

Bayer discards this species as a pentaploid hybrid of unknown parentage. It is of no particular interest to collectors and is subject to brown patches appearing on the leaves, which disfigure the plant.

H. revendettii

H. rigida (Lam.) Haw., *Revis. Pl. Succ.* 49 (1821); *Encycl.* 1:89 (1797), Lamarck—as *Aloe cylindrica* var. *rigida*; *Haw. Handb.* 150 (1976)

Bayer discards this species, similar in appearance to *H. tortuosa*, but on its way to the longer leaves of *H. glabrata*, as of hybrid origin.

H. rossouwii von Poelln., *Kakteenk.* 75 (1938); *Cact. Amer.* 25:8 (1953), J. R. Brown; *Aloe* 12:96 (1974), Bayer; *Haw. Handb.* 151 (1976), Bayer; *Excelsa* 7:39 (1977), Bayer

Bayer rejects this name because of its uncertain identification, and lack of a good type plant. He believes it referable to *H. mirabilis*.

H. rubrobrunnea von Poelln., *Fedde's Repert. Spec. Nov.* 49:57 (1940); *Mem. Soc. Brot.* 2:91 (1943), Resende; *Cact. Amer.* 24:10 (1952), J. R. Brown; *Haw. Handb.* 151 (1976), Bayer

Another hybrid origin plant in the same category as *H. revendettii* and *H. sampaiana*, but with narrower leaves.

H. rubrobrunnea

H. rugosa (Salm-Dyck) Bak., *Journ. Linn. Soc.* 18:206 (1880); Hort. Dyck 323 (1834), Salm-Dyck; *Monogr.* 6:9 (1836–63), Salm-Dyck; *Fl. Cap.* 343 (1896), Dyer; *Cact. Amer.* 37:37 (1965), J. R. Brown; *Haw. Handb.* 151 (1976), Bayer

This old-standing species seems to be intermediate between *H. attenuata* and *H. radula*, with the overall appearance of the latter, but more the tuberculation of the former. Its origins are undetermined (it was described from a plant in the collection of the imperial gardens at Vienna), and its existence separately in the wild is doubted. It is not significantly different from *H. radula* to warrant its retention as a cultivar. *H. rugosa* var. *perviridis* was merely a variant with less white tubercles.

H. rycroftiana Bayer, *Journ. S.A. Bot.* 47:795 (1981); *New Haw. Handb.* 54 (1982), Bayer
Section Retusae subsection Turgidae

This recently described new species is as yet unillustrated, except in Bayer's *New Haworthia Handbook*. It is close to *H. turgida*, apparently representing a link between that species and *H. aristata*. It has triangular, thickish leaves, light green, tapering to an end-awn. It grows vigorously and clusters from around the base, with rosettes each about 6 or 7cm in diameter. It has been known in cultivation for some time, variously labelled, usually as *H. aristata* or *H. unicolor*.

It is reported from near the Gouritz river, between Van Wyksdorp and Herbertsdale.

H. rycroftiana

H. ryderana von Poelln., *Des. Pl. Life* 9:103 (1937); *Fedde's Repert. Spec. Nov.* 43:106 (1938), von Poellnitz; *Haw. Handb.* 151 (1976), Bayer

This is a robust plant probably of hybrid origin, sent to Kew by a Mrs Ryder, and thence to von Poellnitz, who with characteristic lack of reserve promptly described it as a new species. It is still in cultivation in England, but is of no merit when compared with similar true species in the Section Retusae, and not worth retention as a cultivar, the only recognition it could really receive.

H. ryderana

century ago, and by Berger in *Das Pflanzenreich* in 1908. Both these illustrations equate rather with the later-described *H. morrisiae*, which Bayer reduced to a variety of *H. scabra*, maintaining that *H. scabra* was a much more tuberculate species which equated with the later described *H. turberculata*, under which name the more heavily tuberculate form is more often seen in collections. Scott takes the opposite view and upholds *H. tuberculata*.

If Bayer is correct the species resolves as follows:

H. scabra var. *scabra* (syn. *H. tuberculata*).

H. scabra var. *morrisiae* (von Poelln.) Bayer (syn. *H. morrisiae*).

H. sampaiana

H. sampaiana Res., *Bol. Soc. Brot.* 14:192 (1940); *Mem. Soc. Brot.* 2:73 (1943), Resende; *Cact. Amer.* 24:6 (1952), J. R. Brown; *The First Fifty Haw.* 34 (1970), Pilbeam; *Haw. Handb.* 177 (1976), Bayer

This and the form, *H. sampaiana* fa. *broterana* (Res.) Res. & Pinto-Lopes (*Bol. Soc. Brot.* 15:159 (1941) & *Mem. Soc. Brot.* 2:16 (1943)), are dismissed by Bayer as of unknown origin, probably garden hybrids. Like *H. revendettii*, which is similar in appearance, *H. sampaiana* holds no particular place in collectors' hearts, and is not upheld.

H. scabra Haw., *Suppl. Pl. Succ.* 58 (1819); *Revis. Pl. Succ.* 51 (1821); *Monogr.* 7:1 (1836–63), Salm-Dyck; *Cact. Amer.* 29:43 (1957), J. R. Brown; *The Second Fifty Haw.* 51 (1975), Pilbeam; *Haw. Handb.* 152 (1976), Bayer; *Cact. Amer.* 52:274 (1980), Scott
Section Trifariae subsection Acaules

Here is an interesting, beautiful, obstinately home-sick species, still little known in collections, and when obtained kept with some difficulty. It is pictured admirably in Salm-Dyck's Monograph of over a

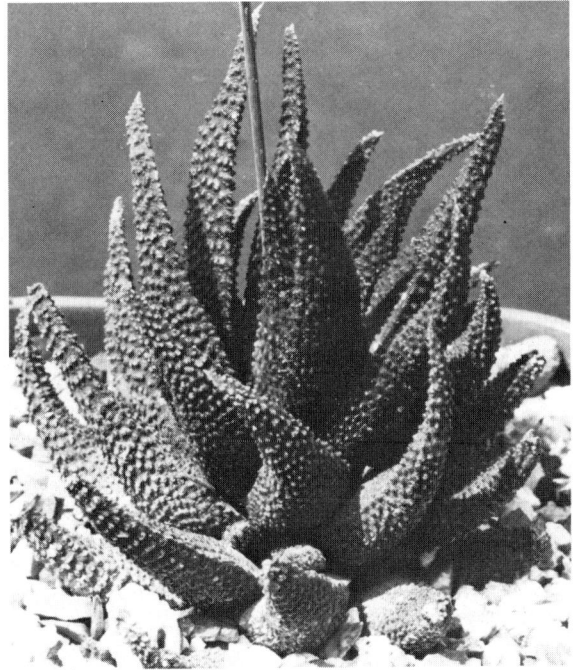

H. scabra var. *scabra*

The type is narrow-leaved with thickly set tubercles the same colour as the leaves, giving the leaf-surface a very rough texture. When growing strongly the leaves open out, but maintain a spiral form, which in periods of dryness becomes more apparent as the leaves close around the stem all curling in the same direction.

H. scabra var. *morrisiae* von Poelln. (Bayer) (*Haw. Handb.* 137 (1976); *Kakteenk.* 132 (1937), von Poellnitz—*H. morrisiae*) tends to grow more strongly in cultivation, with a slightly more open habit, the leaves often narrowing abruptly towards the tip, instead of a steady narrowing as in the type. The tubercles are much smaller, giving the leaves a rough, sandpaper quality, and the overall colouring is a lighter green. Both this and the type, cluster around the base of the rosette, but slowly. Type reported from Ladismith, the variety from Oudtshoorn.

H. scabra var. *morrisiae*

H. scabra var. *morrisiae*

H. schuldtiana von Poelln., *Fedde's Repert. Spec. Nov.* 41:211 (1937); *Des. Pl. Life* 9:101 (1937), von Poellnitz; *Haw Handb.* 152 (1976), Bayer

This species and the many quite superfluous forms described by von Poellnitz in collaboration with Triebner are referred by Bayer to synonymy with *H. magnifica*. A quotation from the text when the many forms were described shows their value: 'From this species we can split off a number of forms which grow with it in a comparatively small region, according to the information in a letter from Triebner, several forms never grow together, but each small hill may produce a different form.' Enough said. The only two singled out for special treatment by Bayer are G. G. Smith's *H. schuldtiana* var. *major*, which he makes a variety of *H. magnifica*, and *H. schuldtiana* var. *maculata*, which he erects to species status.

H. semiglabrata Haw., *Suppl. Pl. Succ.* 55 (1819); *Cact. Amer.* 43:157 (1971), Bayer; *Haw. Handb.* 152 (1976), Bayer

Referred by Bayer to *H. margaritifera*, now *H. pumila*.

H. semiviva (von Poelln.) Bayer, *Haw. Handb.* 153 (1976); *Fedde's Repert. Spec. Nov.* 44:135 (1938), von Poellnitz—*H. bolusii* var. *semiviva*; *Cact. Amer.* 13:87 (1941), J. R. Brown
Section Arachnoideae subsection Limpidae

This former variety of *H. bolusii* has been raised to specific status by Bayer. He maintains that its resemblance to *H. bolusii* is superficial, and that its separate occurrence at Beaufort West indicates more probable relationship to *H. arachnoidea* or *H. lockwoodii*, with which it shares a basic light green coloration. It is characterized by long, translucent bristles on the margins and keels of the leaves, which interlace at random, and by the tendency of the very thin tips of the leaves to die back if the plant is allowed to become at all dry at the root; in fact it is almost impossible to prevent this quite natural occurrence. It does not readily cluster, and once it has attained about 7 or 8cm in diameter it seems to gain and lose about the same number of leaves each year. As with similar species it is susceptible to overmuch water lying around amongst the leaves, which is liable to let rot set in, and with such little substance to the leaves it is difficult to save once it has got a foothold. Propagation from leaves is almost impossible, but it comes easily from seed.

H. semiviva

H. serrata Bayer, *Journ. S.A. Bot.* 39:249 (1973); *The Second Fifty Haw.* 14 (1975), Pilbeam; *Haw. Handb.* 153 (1976), Bayer
Section Retusae subsection Turgidae

This recently described species has not got into collections much as yet. It has many leaves to the rosette, about 50 or more, with prominent teeth on the margins of the leaves. The colouring is usually a bright grass-green, with somewhat translucent, slightly recurved end-areas to the leaves. It comes from an area south of the Langeberg mountain range, 40km southwest of Heidelberg.

H. serrata

H. sessiliflora Bak., *Fl. Cap.* 6:353 (1896) Dyer; *Haw. Handb.* 153 (1976), Bayer

Bayer has been unable to relate this name to field populations, and with its uncertain identity it is a name best discarded. Plants seen under the name in cultivation in England are generally referable to the Section Retusae, and although they are translucent in the end-areas of the leaves it is unlikely that they would be readily ascribed to the old Section Limpidae, which this species was.

H. serrata

H. setata Haw., *Suppl. Pl. Succ.* 52 (1819); *Monogr.* 12:3 (1836–63), Salm-Dyck; *Cact. Amer.* 16.3 (1944), J. R. Brown; & 49:205 (1977), C. L. Scott; *The First Fifty Haw.* 34 (1970), Pilbeam

This well-known name has been sunk into synonymy with the prior *H. arachnoidea*, although not without considerable resistance from Scott, who maintains its separate identity. It is not upheld separately here.

H. skinneri (Berg.) Res., *Mem. Soc. Brot.* 3:76 (1943); *Pflanz.* 4:116 (1908), Berger—*Apicra skinneri*; *Succulenta* 53 (1947), Uitewaal—*Astroloba skinneri*; *Nat. Cact. Succ. Journ.* 28:7 (1973), Rowley— × *Astroworthia bicarinata* nm. *skinneri*

A natural hybrid between *Astroloba aspera* and *H. pumila*. See *Astroloba skinneri*, page 153.

H. smitii

H. smitii von Poelln., *Des. Pl. Life* 10:186 (1938); *Haw. Handb.* 154 (1976), Bayer; *Cact. Amer.* 54:162 (1982). Scott
Section Trifariae, subsection Acaules

This species was recorded from Oudtshoorn. Its exact locality has been unknown for many years, but Scott has recently found it in the Kammanasie mountains in the Oudtshoorn district. Its comparatively thin flower stem indicates Bayer's alliance with *H. starkiana* is nearer the truth than Scott's tentative suggestion that it is near *H. kingiana* in the section Margaritiferae.

H. sordida Haw., *Revis. Pl. Succ.* 51 (1821); *Monogr.* 7:2 (1836–63), Salm-Dyck; *Fedde's Repert. Spec. Nov.* 43:232 (1938), Zantner & von Poellnitz—*H. agavoides*; *Journ. S.A. Bot.* 16:2 (1950), G. G. Smith—*H. sordida* var. *agavoides*; *Cact. Amer.* 17:51 (1945), J. R. Brown; & 53:124 (1981), Scott—*H. sordida* var. *lavrani*; *The First Fifty Haw.* 36 (1970), Pilbeam; *Haw. Handb.* 155 (1976), Bayer
Section Trifariae subsection Acaules

H. sordida

This delightful, difficult species is the *Neogomesia* of Haworthias, coming from arid, inhospitable terrain (aptly named Dead Man's Gulch at one location), and resisting attempts to grow it rapidly in cultivation. It has oddly rounded ends to the leaves and a rough texture due to self-coloured tubercles scattered at random all over the surface of the leaves, but sometimes forming into disorderly ranks across the width. The leaf is thick and has wide, flattened margins with shallow excavated centres in the upper side, with little evidence of a keel. It is reported from the Uitenhage area and to the north-east near Steytlerville and Jansenville.

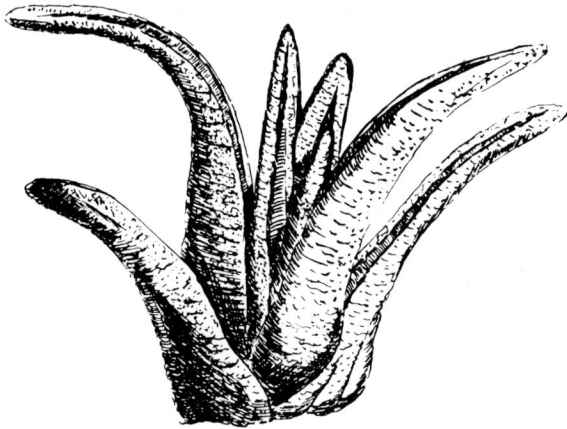

H. sordida

H. sordida var. *agavoides* (Zant. & von Poelln.) G. G. Smith was reported by Smith to occur near Maraishope and Kleinpoort in the Uitenhage district. It differed from the type according to Smith by the distinct tubercles compared with the tuberculate-rugose surface of the type. Zantner and von Poellnitz named it *H. agavoides* for its differently formed leaves, i.e. like an Agave. Which leads me to the opinion that the recently described *H. sordida* var. *lavrani* C. L. Scott from a restricted area—Perde Hoek farm, 36km south-east of Steytlerville, is a redescription of this variety. Whatever you call it the differences from the type are minimal and hardly warrant varietal status. A light position is needed to keep the blackish-green coloration of this species, and a light hand with water, allowing the compost to dry out well between waterings. Give room too for the thick, searching roots. If grown well it has a dark reptilian beauty reminiscent of such other succulent delights as *Ceropegia armandii* and *Gasteria batesiana*.

H. springbokvlakensis C. L. Scott, *Journ. S.A. Bot.* 36:287 (1970); *Aloe* 11:39 (1973), C. L. Scott; *Excelsa* 5:83 (1975), Bayer; *The Second Fifty Haw.* 12 (1975), Pilbeam; *Haw. Handb.* 155 (1976), Bayer
Section Retusae subsection Retusae

This species has already become, in the short time it has been in cultivation, one of those in the genus most sought after by collectors. It is indeed a real gem, with the best attributes of this desirable section. The almost

wholly translucent end-areas of the leaves are pulled down to soil-level in the manner of *H. truncata* or *H. maughanii*, the rounded ends to the leaves fitting tightly together to protect the submerged body of the plant, recurving strongly. It comes from the farm Springbokvlakte in the Uitenhage district, east of Steytlerville in exposed positions growing level or partly submerged in the sandy soil, and completely submerging in the hot summer months. Caution in cultivation is necessary, with a deep enough pot to allow for the contraction into the compost which invariably takes place, and careful watering at all times; make haste slowly.

H. springbokvlakensis

H. springbokvlakensis

H. starkiana von Poelln., *Fedde's Repert. Spec. Nov.* 33:73 (1933); *Des. Pl. Life* 7:132 (1935), von Poellnitz; *Cact. Amer.* 15:176 (1943), J. R. Brown; *The First Fifty Haw.* 37 (1970), Pilbeam; *The Second Fifty Haw.* 36 (1975), Pilbeam; *Haw. Handb.* 128 & 155 (1976), Bayer
Section Trifariae subsection Acaules

Bayer placed the former species *H. lateganiae* at varietal level under this species in his Handbook.

The type, *H. starkiana* var. *starkiana*, has smooth, three-angled leaves, with slightly cartilaginous margins. It makes a fairly squat rosette of short leaves (about 4 or 5cm long), wide-spreading at a low angle, becoming yellowish in full light. It comes from Schoemanspoort in the Oudtshoorn, virtually from within the town of Peddie, and does not seem to exist there any more.

H. starkiana var. *lateganiae*

H. starkiana var. *starkiana*

H. starkiana var. *starkiana*

H. starkiana var. *lateganiae*

H. starkiana var. *lateganiae* (von Poelln.) Bayer, *Des. Pl. Life* 9:103 (1937), von Poellnitz; *Fedde's Repert. Spec. Nov.* 43:99 (1938), von Poellnitz; *Cact. Amer.* 15:176 (1943), J. R. Brown; *The First Fifty Haw.* 37 (1970), Pilbeam; *The Second Fifty Haw.* 36 (1975), Pilbeam; *Haw. Handb.* 128 (1976), Bayer. This variety comes from a neighbouring valley to the east of the type, Klein Kruis. In either variety this is a difficult species to grow well, and in this variety, which differs only in its longer leaves, up to about 10cm long, and its taller habit, forming a stem to about 5cm tall in time, compared with the almost stemless habit of the type, it is difficult to keep the tight spiralling habit, which to my way of thinking makes for the most attractive way of growing this variety. To retain this sort of growth a light position should be given and a light hand with the watering, when the light green colouring can become almost orange.

H. stiemei von Poelln., *Fedde's Repert. Spec. Nov.* 44:227 (1938); *Haw. Handb.* 156 (1976), Bayer

There is no clear idea as to the identity of this species erected by von Poellnitz just before World War 2 and unknown in collections since. Apparently the Mr Stiemie for whom it was named is unaware of such a species or of having collected such a plant. Bayer expresses the tentative opinion that it may be a variant of either *H. translucens* or *H. xiphiophylla*, which occur in areas neighbouring Kirkwood, where this species was reported to have been collected, but the reliability of von Poellnitz's reported localities is not high, and the species remains shrouded in doubt, and is best discarded as insufficiently known. Plants in cultivation under this name in England are usually referable to *H. cooperi* fa. *pilifera*.

H. cv. subattenuata (Salm-Dyck) Bak., *Hort. Dyck.* 324 (1834), Salm-Dyck; *Monogr.* 6:11 (1836–63),

H. cv. *subattenuata*

H. cv. *subattenuata* (variegated form)

Salm-Dyck; *Journ. Linn. Soc.* 18:205 (1880), Baker; *Cact. Amer.* 14:132 (1942); & 18:119 (1946), J. R. Brown; & 43:157 (1971), Bayer; *ASPS* 3:194 (1968), Parr & Pilbeam; *Haw. Handb.* 156 (1976), Bayer

This species was discarded by Bayer in his article in the American society's journal (1971 ref. above), although in his Handbook (1976) he leaves it in limbo, with no note of explanation at all. He doubted that it could be referred to field populations. However, plants in collections today are well-matched with Salm-Dyck's illustrations of over a century ago, with almost smooth upper surfaces to the large, triangular in section leaves, and fairly evenly distributed small tubercles on the backs of the leaves. It is more commonly seen as a variegated plant, with large areas of yellow, chlorophyll-less epidermis, usually in longitudinal stripes, but occasionally with completely yellow leaves. It may represent a hybrid between *H. pumila* and *H. attenuata* at a guess, but whatever its origins it is worth retaining as a cultivar, because of its persistent popularity in collections and its attractive appearance, especially in the variegated form.

H. subfasciata (Salm-Dyck) Bak., *Journ. Linn. Soc.* 18:204 (1880), Baker; *Hort. Dyck.* 325 (1834), Salm-Dyck; *Monogr.* 6:15b (1836–63), Salm-Dyck; *Cact.* *Amer.* 38:4 (1966), J. R. Brown; & 43:157 (1971), Bayer; *Haw. Handb.* 157 (1976), Bayer

Similarly to *H. subattenuata* Bayer in his American society's journal article (1971 ref. above) doubted that this species equated with field populations and raised the former variety, *H. subfasciata* var. *kingiana* von Poelln. (*Fedde's Repert. Spec. Nov.* 41:203 (1937) & *loc. cit*: 218 (1938), to its former species status. Again in his Handbook (1976) Bayer makes no comment under *H. subfasciata*. The plant pictured in Salm-Dyck's Monograph seems to be something akin to *H. fasciata* var. *browniana*. It is not a name applied to plants in cultivation today, except in error, and is best discarded, since it cannot be applied with certainty.

H. sublimpidula von Poelln., *Fedde's Repert. Spec. Nov.* 41:212 (1937); *Cact. Journ.* 5:33 (1936), von Poellnitz; *Cact. Amer.* 26:74 (1954), J. R. Brown; *Cact. Succ. Journ. NSW.* 8:9 (1971), Bayer; *Haw. Handb.* 158 (1976), Bayer

Bayer in his Handbook (1976) refers this species to synonymy with *H. magnifica*. There is a lingering suspicion in my mind that in describing *H. magnifica* var. *meiringii* (corrected from *H. maraisii* var. *meiringii* by Bayer in *Nat. Cact. Succ. Journ.* 32:18 (1977) Bayer was redescribing this species, *H. sublimpidula*.

H. cv. *subattenuata* (variegated form, with almost completely chlorophyll-less leaves in the new growth)

H. submaculata von Poelln., *Des. Pl. Life* 11:194 (1939); *Cact. Amer.* 14:10 (1942), J. R. Brown; *Cact. Succ. Journ. NSW.* 8:9 (1971), Bayer; *Nat. Cact. Succ. Journ.* 27:51 (1972), Bayer; *The Second Fifty Haw.* 6 (1975), Pilbeam; *Haw. Handb.* 159 (1976), Bayer

Bayer refers this plant to synonymy with *H. herbacea*; it is referable to a hybrid swarm 9 miles south of Worcester between *H. herbacea* and *H. maculata*.

H. subregularis Bak., *Saund. Refug. Bot.* 4:t.282 (1870); *Nat. Cact. Succ. Journ.* 27:10 (1972), Bayer; *The Second Fifty Haw.* 6 (1975), Pilbeam; *Haw. Handb.* 159 (1976), Bayer

Bayer refers this old name to synonymy with *H. reticulata*, where it is maintained herein to represent the larger forms of this species at *forma* level.

H. subulata (Salm-Dyck) Bak., *Journ. Linn. Soc.* 18:206 (1880); *Hort. Dyck.* 324 (1834), Salm-Dyck; *Haw. Handb.* 159 (1976), Bayer

Bayer declares this species to be probably a garden hybrid. Plants in cultivation seem to be a smaller variant of *H. radula*.

H. tauteae Arch., *Fl. Pl. Afr.* 25:t.992 (1946); *Haw. Handb.* 160 (1976) Bayer

Bayer says that this species is a natural hybrid between *H. viscosa* and *H. scabra*, occasionally found at Kleinspoort, east of Oudtshoorn. It is sometimes seen in collections labelled as *H. tortuosa* var. *tortella*, with which it may well be synonymous, but since *H. tortuosa* as a species is rejected by Bayer as unrelated to field populations, the point is academic.

H. tenera von Poelln., *Fedde's Repert. Spec. Nov.* 31:86 (1932); *Journ. Linn. Soc.* 18:215 (1880), Baker— as *H. minima*, an invalid name through prior use.

Bayer refers this species to subspecific status beneath *H. translucens*.

H. tessellata Haw., *Phil. Mag.* 44:300 (1824); *Monogr.* 8:1 (1836–63) Salm-Dyck; *The Second Fifty Haw.* 53 (1975), Pilbeam; *Haw. Handb.* 161 (1976), Bayer

I was reluctant to accept Bayer's submerging of this species, so well-known and aptly named, in favour of the prior names *H. venosa* and *H. recurva*, the latter of which was declared synonymous with *H. tessellata* and reduced to subspecific status beneath *H. venosa*. The resurrection of such old names in preference for commonly accepted ones tends to bring the impatience of collectors with botanists to the boil. More so now as Bayer, in his *New Haworthia Handbook* (1982), has reinstated the name 'tessellata' over 'recurva'. But let us not question this change in the wind when it blows favourably in our direction—see under *H. venosa* subsp. *tessellata*.

H. tauteae

H. tisleyi Bak., *Journ. Linn. Soc.* 18:208 (1880); *Fl. Cap.* 6:344 (1896), Dyer; *Cact. Amer.* 43:157 (1971), Bayer; *Haw. Handb.* 160 (1976), Bayer

Its origin unknown, this nondescript species is regarded by Bayer as probably of garden origin. Plants seen in collections under this name have affinities with *H. glabrata* or *H. tortuosa*, and are of no appeal to the collector. Their apparent hybrid origins are underlined by their apparent inability to flower.

H. tortuosa Haw., *Syn. Pl. Succ.* 90 (1812); *Monogr.* 4:2 (1836–63); *Fl. Cap.* 6:336 (1896), Dyer; *Cact. Amer.* 14:26 (1942), J. R. Brown; *Nat. Cact. Succ. Journ.* 24:41 (1969), Pilbeam; *The First Fifty Haw.* 38 (1970). Pilbeam.

This species in its various forms is a well-known member of many collections, with dark green, tuberculately rough, three-sided leaves, sharply pointed and markedly spiralling. The length of the leaf varies from about 2.5cm long (*H. tortuosa* var. *curta* Haw. to 4 or 5cm long in the type. This and the other varieties (*H. tortuosa* var. *pseudorigida* (Salm-Dyck) Berg., *H. tortuosa* var. *tortella* (Haw.) Bak., and *H. tortuosa* var.

this species, which may give an indication of its origins. There is an attractive variegated form.

H. translucens Haw., *Syn. Pl. Succ.* 52 (1812); *Trans. Linn. Soc.* 7:10 (1804), Haworth—*Aloe translucens; Haw. Handb.* 162 (1976), Bayer; *New Haw. Handb.* 56 (1982), Bayer
Section Arachnoideae subsection Limpidae

H. translucens subsp. *translucens*

With this species Bayer couples the better known name *H. tenera*. It is a dainty, thin-leaved, heavily bristled species, forming clusters of rosettes no more than about 5cm in diameter. The leaves are translucent, more so at the tips, narrowly lanceolate, and ending in a fine, bristled awn, and numerous, 50 or more to each rosette, clustering from around the base to form small clusters, grey-green, becoming lilac-grey in full light.

H. tortuosa

H. translucens subsp. *translucens*

H. tortuosa (variegated form)

major (Salm-Dyck) Berg.) can be discarded along with the type as not relating to field populations according to Bayer. Certainly none of them have ever been described from wild collected plants, and none has ever been reported from the wild in all the time they have been known in cultivation. Hybrids produced between *H. viscosa* and other species in cultivation resemble

H. translucens subsp. *tenera* is altogether smaller, and more intolerant of excessive sunlight than is the type.

Reported from the Gamtoos Valley west of Port Elizabeth; subsp. *tenera* from east of Grahamstown.

H. truncata fa. *truncata*

H. truncata fa. *crassa*

H. translucens subsp. *tenera*

This well-known species beautifully adapted for survival, has leaves which have the appearance (as the name indicates) of having had the ends scythed off. In habitat its thick, contractile roots pull it down in times of drought so that it is level or even below the soil surface, leaving only the translucent ends of the leaves to allow the light through to the inner parts of the leaves to carry out the process of photosynthesis necessary to allow the plant to grow in these Spartan circumstances. There is an obvious parallel here with the 'window-leaved' Mesmybryanthemaceae, such as *Lithops, Fenestraria, Frithia* etc. Karl von Poellnitz split the species into three forms, according to the thickness of the end of the truncated leaves, viz. fa. *truncata* with leaf-ends 6 to 8mm thick, fa. *tenuis* with leaf-ends 3 to 4mm thick, and fa. *crassa*, with leaf-ends

H. translucens subsp. *tenera*

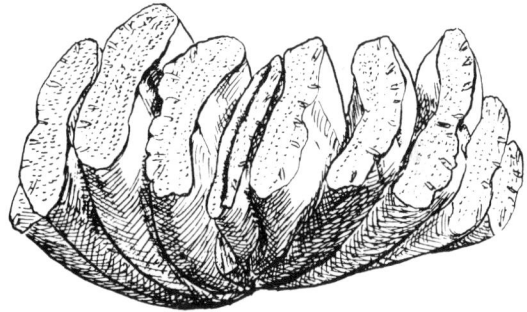

H. truncata fa. *truncata*

H. triebnerana von Poelln., *Fedde's Repert. Spec. Nov.* 41:214 (1937); & 47:8 (1939), von Poellnitz; *Cact. Journ.* 5:33 (1936); & 6:36 (1937) von Poellnitz; *Aloe* 11:23 (1973), C. L. Scott; *Excelsa* 5:87 (1975), Bayer; *Haw. Handb.* 163 (1976), Bayer; *Excelsa* 7:37 (1977), Bayer

It was difficult to face the welter of varieties that von Poellnitz set up for this species, without feeling that they could be reduced to many fewer, and Bayer did this very effectively in his Handbook (1976), reducing the whole species to synonymy with *H. mirabilis*. However some varieties erected by von Poellnitz are striking forms, and as such are retained beneath *H. mirabilis*.

H truncata Schönl. apud Marl, *Berl. Deutsch. Bot. Ges.* 27:367 (1909); *Trans. Roy. Soc. S.A.* 1:391 (1910), Schönland; *Cact. Amer.* 9:86 (1937); & 16:86 (1944); & 27:35 (1955), J. R. Brown; *Fedde's Repert. Spec. Nov.* 44:237 (1938), von Poellnitz; *Des. Pl. Life* 15:151 (1943), J. R. Brown; *The First Fifty Haw.* 39 (1970), Pilbeam; *The Second Fifty Haw.* 53 (1975), Pilbeam; *Haw. Handb.* 163 (1976), Bayer
Section Fenestratae

9 to 11mm thick. Bayer maintained only the narrow-leaved form in his 1976 Handbook, raising it to varietal level because of its separation in habitat, and merging fa. *crassa* with the type since they occur in the same locality. He subsequently reduced both to synonymy in his 1982 Handbook, since the narrow var. *tenuis* has been observed not necessarily to maintain its narrowness of leaf-end in cultivation—not this writer's experience. Because of the popularity of the species, especially of the differing forms, all three are maintained at form level here.

The species resolves as follows:
H. truncata Schönl. fa. *truncata.*
H. truncata fa. *crassa* von Poelln.
H. truncata fa. *tenuis* von Poelln.

Some plants remain solitary in cultivation reaching a maximum number of leaves of about 12, losing as many each year from the outer limits of the fan-shaped spread of leaves as they gain from new ones produced in the centre. Others cluster quickly to make in time superb clumps of many heads, ranging this way and that in a crazy mosaic. If grown hard, so that little of the lower parts of the leaves can be seen between the truncate tops, these clumps look magnificent, but too often I have seen them growing open and pale green which is not to make the best of the species.

H. truncata fa. *tenuis*

H. truncata × *H. limifolia*

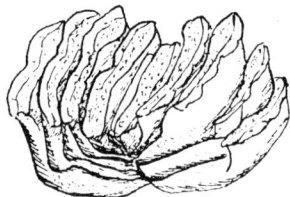

H. truncata fa. *tenuis*

H. tuberculata von Poelln., *Fedde's Repert. Spec. Nov.* 29:219 (1931); *Haw. Handb.* 163 (1976), Bayer

Bayer takes *H. scabra* to be synonymous with this species, and therefore by priority the preferred name; see under *H. scabra*.

H. turgida Haw., *Suppl. Pl. Succ.* 52 (1819); *Monogr.* 9:5 (1836–63), Salm-Dyck; *Des. Pl. Life* 8:29 (1936), J. R. Brown; *Cact. Amer.* 29:133 (1957), J. R. Brown; *Aloe* 11:37 (1973), Scott; *Excelsa* 5:89 (1975), Bayer; *Haw. Handb.* 163 (1976), Bayer
Section Retusae subsection Turgidae

Bayer merges *H. laetevirens* and *H. caespitosa* with this species and recognizes none of the varieties (var.

pallidifolia, var. *suberecta* and var. *subtuberculata*). He says that it is a widely spread species particularly in the sandstone mountains of the Langeberg and Rivierson-derend ranges. It varies ecotypically where it occurs in shade and differing soils, and abuts (and possibly hybridizes with) no less than four other species, viz. *H. retusa*, *H. reticulata*, *H. mirabilis* and *H. herbacea*.

This small, turgid-leaved species is popular in collections and the illustrations give an idea of the variation. It presents no difficulty in cultivation, being tolerant of under or over-watering, within reason, and sun or shade, but to look its best sufficient exposure to the sun, to bring out reddish hues and keep the leaves close together in a compact rosette or a compact cluster of rosettes, is the best treatment.

H. turgida fa. *turgida*

H. turgida fa. *turgida*

Some recognition of this widely variable species would not be amiss for the benefit of collectors, and the former varieties *H. turgida* var. *pallidifolia* and *H. turgida* var. *suberecta* (the latter embracing *H. turgida* var. *subtuberculata*) are upheld at form level, as well as the former species *H. caespitosa*.

This species resolves as follows:

H. turgida Haw. fa. *turgida* (syn. *H. laetevirens*)

H. turgida fa. *caespitosa* (von Poelln.), Pilbeam stat. nov. (syn. *H. caespitosa* von Poelln., *Fedde's Repert. Spec. Nov.* 43:103 (1938)).

H. turgida fa. *pallidifolia* (G. G. Smith) Pilbeam stat. nov. (syn. *H. turgida* var. *pallidifolia* G. G. Smith, *Journ. S.A. Bot.* 12:10 (1946)).

H. turgida fa. *suberecta* von Poelln.) Pilbeam stat. nov. (syn. *H. turgida* var. *suberecta* von Poelln.—*Fedde's Repert. Spec. Nov.* 44:134 (1938)—and *H. turgida* var. *subtuberculata* von Poelln.—*Fedde's Repert. Spec. Nov.* 44:134. (1938)).

H. turgida fa. *caespitosa*

H. turgida fa. *caespitosa* makes small rosettes, much thinner-leaved than other forms and with prominent teeth on the margins of the leaves; it has a flattish growing habit, with fewer leaves to the rosette than other forms, and colours well to reddish hues in full light. Although reported from McGregor, Kleinpoort (Oudtshoorn) and Uitenhage, Bayer gives its most probable origins as the Tradouw Pass, east of Swellendam.

Bayer reports that *H. turgida* fa. *pallidifolia* is recorded from north of Albertinia where it grows along the Valsch river in dense clusters. It is probably the most widely grown form of this species in cultivation, and is most attractive if grown well, making flattish mounds of rosettes with grey-green, semi-translucent end-areas of the leaves prominently marked with whitish flecks, which give a frosted appearance, although the flecks are beneath the surface.

H. turgida fa. *suberecta* makes a cluster of small rosettes with leaves extending upwards in a more erect habit than the type or other forms, so that the clumps formed are as tall as wide. This form tends to redden in full light more than the previous form and the type. It is reported from George in the Karoo, although Bayer discounts this locality in favour of Brandwacht near Mossel Bay.

H. turgida fa. *suberecta*

H. ubomboensis Verdoorn, *Fl. Pl. S. Afr.* 21:t.818 (1941); *Journ. S.A. Bot.* 16:3 (1950), G. G. Smith—*H. limifolia* var. *ubomboensis*; *The Second Fifty Haw.* 39 (1975), Pilbeam; *Haw. Handb.* 164 (1976), Bayer Section Venosae

H. turgida fa. *pallidifolia*

H. ubomboensis

This species is returned to its specific status, since cytological examination by Dr Peter Brandham of the Jodrell Laboratory, Royal Botanic Gardens, Kew, indicates that it cannot be a variety of *H. limifolia* (unpublished findings).

Although the general habit of *H. ubomboensis* is similar to *H. limifolia*, it is a much less rigid-leaved species, and with no indication of the ridges characteristic of the latter. The colouring too, compared with the dark green of *H. limifolia*, is a pale green turning to lilac shades in full light. It offsets stoloniferously in the same way as *H. limifolia*, however, but presents more difficulty in cultivation, appearing to be susceptible to overwatering, when it loses its roots. It is reported from Swaziland, east of Stegi in the Ubombo mountains, but has not been recollected recently.

H. ubomboensis

H. uitewaaliana von Poelln., *Cact. Vetpl.* 5:115 (1939); *Cact. Amer.* 33:86 (1961), J. R. Brown; & 43:157 (1971), Bayer; *Haw. Handb.* 164 (1976), Bayer

This species has disappeared under Bayer's rationalization of the subgenus Robustipedunculares, declared synonymous with *H. marginata*, of which it is merely a more tuberculate form, or perhaps a hybrid between *H. marginata* and *H. minima*.

H. umbraticola von Poelln., *Kakteenk.* 134 (1937); *Des. Pl. Life* 9:103 (1937), von Poellnitz; *Cact. Amer.* 17:3 (1945), J. R. Brown; *The Second Fifty Haw.* 56 (1975), Pilbeam; *Haw. Handb.* 164 (1976), Bayer

Bayer reduces this species to varietal status beneath *H. cymbiformis*.

H. unicolor von Poelln., *Kakteenk.* 154 (1937); *Haw. Handb.* 165 (1976), Bayer; *Nat. Cact. Succ. Journ.* 35:11 (1980), Scott; *New Haw. Handb.* 58 (1982), Bayer

With the uncertainty surrounding the application of the name *H. aristata*, Bayer in his Handbook (1976) reinstated this species erected in the 1930s by von Poellnitz, but submerged subsequently beneath *H. aristata*. He also reduced the species *H. helmiae* and *H. venteri* to varietal status beneath *H. unicolor*. Scott disagrees with Bayer on the basis of an illustration in the Herbarium Library at the Royal Botanic Garden, Kew, and examination of that illustration, which is contemporary with Haworth's erection of *H. aristata*,

confirms Scott's view that the name *H. aristata* should take preference for those species known under the names *H. unicolor*, *H. helmiae* and *H. venteri*. See under *H. aristata*.

H. variegata L. Bol., *Journ. Bot.* 67:137 (1929); *Fedde's Repert. Spec. Nov.* 28:102 (1930), von Poellnitz; *Cact. Amer.* 12:129 (1940), J. R. Brown; *Cact. Succ. Journ. NSW.* 9:66 (1974), Bayer, *The Second Fifty Haw.* 56 (1975), Pilbeam; *Haw. Handb.* 165 (1976), Bayer
Section Loratae subsection Loratae

This species has wonderfully marked leaves, with translucent flecks pervading the whole surface area, and a multitude of small translucent teeth. The leaves are long and attenuated, not stiff, and tend to curve this way and that as though wind-blown, the outer leaves recurving strongly to lay almost flat, the inner ones upright and clasping together reluctant to part company. It offsets from around the base, and eventually makes a small clump of rosettes each of about 30 leaves and no more than 3 or 4cm in diameter. In shade it will be dark green with paler markings where the flecking occurs; in full light it will become almost brown, but the markings are less contrasting. It is reported from a coastal area between Riversdale and Stilbaai.

H. variegata

H. variegata

H. venosa subsp. *venosa* in the wild, by the side of the Breede river in brown shale

H. venosa subsp. *granulata*

leaves up to 12cm long, but towards the southern part of its range, near Napkysmond, smaller forms are found with leaves only 2 or 3cm long.

H. venosa subsp. *granulata* merges with the previous subspecies, coming from several parts of the Ceres Karoo. At its most distinguishable it is columnar in growth, and the backs of the leaves are heavily tuberculate, but less caulescent forms are found, and the division from other subspecies becomes blurred. This and some of the more slow-growing forms of the previous subspecies sometimes present difficulties in cultivation, and an open compost and careful watering are called for, but most forms of the species grow well and make large clusters, offsetting stoloniferously.

H. venosa subsp. *tessellata* has probably the widest distribution of any *Haworthia* according to Bayer, 'throughout the Cape Province, in South West Africa and in the Orange Free State', and the size too varies enormously, as well as the degree of translucence and tuberculation. Its constant characteristic is usually low, short-stemmed growth, with strongly recurving leaves, the upper surface semi-translucent with opaque criss-cross lines in tessellate fashion.

H. venosa (Lam.) Haw., *Revis. Pl. Succ.* 51 (1821), *Encycl.* 1:89 (1783), Lamarck; *Des. Pl. Life* 7:61 (1935), J. R. Brown; *Haw. Handb.* 166 (1976), Bayer; *Cact. Amer.* 50:74 (1978), Scott; *New Haw. Handb.* 76 (1982), Bayer
Section Venosae

Bayer concluded in his Handbook (1976) that this species embraced *H. granulata*, which he reduced to subspecific level, and *H. recurva*, which he took as the prior name embracing *H. recurva* and *H. tessellata*, maintaining this name at subspecific level too. In his handbook revision (*The New Haworthia Handbook*, 1982) however he bows to Scott's preference to retain the name *H. tessellata*, and replaces *H. venosa* subsp. *recurva* with *H. venosa* subsp. *tessellata*. *H. venosa* var. *oertendahlii* Hjelmquist (*Botanska Notiser.* 233 (1943)) is not upheld by Bayer, and likewise he dismisses the varieties of *H. tessellata*, mostly created by Resende on the basis of a selection of plants received from various sources unrelated to field populations. As Bayer has stated in correspondence with the author, 'there are more attractive forms of *H. tessellata* in the field than you could put names to'.

H. venosa (Lam.) Haw. subsp. *venosa*.

H. venosa subsp. *granulata* (Marl.) Bayer (syn. *H. granulata*).

H. venosa subsp. *tessellata* (Haw.) Bayer (syn. *H. recurva, H. tessellata, H. pseudotessellata*).

H. venosa subsp. *venosa* occurs in the Lower Breede river valley from Swellendam southwards. At the northern part of its distribution it is large, with erect

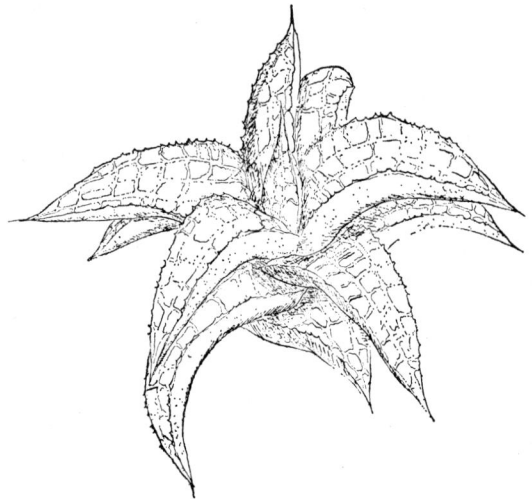

H. venosa subsp. *tessellata*

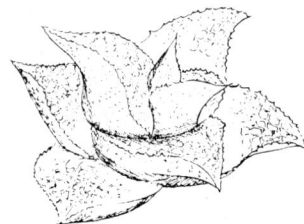

H. venosa subsp. *tessellata (parva)*

H. venosa subsp. *tessellata*

H. venosa subsp. *tessellata*

H. venosa subsp. *tessellata*

H. venosa subsp. *tessellata*

H. venosa subsp. *tessellata*

H. venosa subsp. *tessellata*

H. venteri von Poelln., *Cact. Journ.* 8:19 (1939); *Journ. S.A. Bot.* 14:55 (1948), G. G. Smith; *Cact. Amer.* 36:74 (1964), J. R. Brown; *Haw. Handb.* 166 (1976), Bayer; *Nat. Cact. Succ. Journ.* 35:11 (1980), Scott; *New Haw. Handb.* 59 (1982), Bayer

There is no doubt, after examination of the contemporary illustration of *H. aristata* in the Herbarium Library at Kew, that *H. venteri* is synonymous with that species. See under *H. aristata*.

H. viscosa (L.) Haw., *Syn. Pl. Succ.* 90 (1812); *Sp. Pl.* 460 (1753), Linnaeus; *Pl. Succ. Hort. Alenc.* 7 (1809), Duval; *The Second Fifty Haw.* 58 (1975), Pilbeam; *Haw. Handb.* 167 (1976), Bayer; *Nat. Cact. Succ. Journ.* 36:98 (1981), Scott
Section Trifariae subsection Caulescentes

This deservedly popular species has been in cultivation for a long time, and has gathered some varietal names in the course of time, most of which are hardly warranted on the fine differences they represent. It has lost, or rather never found, a few varieties which have been set up as separate species, i.e. *H. cordifolia, H. asperiuscula* and *H. beanii*.

Taking first the varieties ascribed to the species we have *H. viscosa* var. *caespitosa* von Poelln. (*Fedde's Repert. Spec. Nov.* 44:240 (1938); *Des. Pl. Life* 11:8 (1939); *Cact. Amer.* 14:36 (1942), J. R. Brown; *The Second Fifty Haw.* 58 (1975), Pilbeam), described as coming from Vensterkrans near Ladismith, which, although named for its clumping habit, differs principally from other varieties in its shorter stems, about 6cm, since all varieties clump more or less rapidly in time; this variety is in the sprint class compared with others, making offshoots readily. *H. viscosa* var. *cougaensis* G. G. Smith (*Journ. S.A. Bot.* 11:65 (1945)), not 'coegaensis' as in Jacobsen's *Handbook of Succulent Plants*, was described from the Willowmore Division between Couga and Zandvlakte; it differs little from the type. *H. viscosa* var. *concinna* (Haw.) Bak. (*Suppl. Pl. Succ.* 59 (1819); *Monogr.* 3:4 (1836–63), Salm-Dyck; *Journ. Linn. Soc.* 18:200 (1880), Baker), has straight rows of leaves, not slowly twisting as in the type, nor sharply twisting or as long-leaved as in var. *torquata*, not really a distinction warranting varietal recognition. *H. viscosa* var. *indurata* (Haw.) Bak. (*Revis. Pl. Succ.* 49 (1821); *Journ.*

H. viscosa fa. *asperiuscula*

H. viscosa fa. *beanii*

H. viscosa fa. *asperiuscula*

Haworth; *Monogr.* 3:5 (1836–63), Salm-Dyck; *Journ. Linn. Soc.* 18:201 (1880), Baker; *Cact. Amer.* 14:83 (1942), J. R. Brown; *The First Fifty Haw.* 39 (1970), Pilbeam), is distinctly different in appearance, spiralling and with leaves more rounded in the lower half, suddenly tapering to a fine point. It swiftly (for this species) make clumps from around the base, filling a 12cm pan in about five or six years. *H. viscosa* var. *quaggaensis* G. G. Smith (*Journ. S.A. Bot.* 14:46 (1948), makes more compact, closer-leaved stems than the type, according to the description by Smith, but differs little from the following variety. *H. viscosa* var. *subobtusa* von Poelln. (*Fedde's Repert. Spec. Nov.* 44:240 (1938); *Des. Pl. Life* 11:8 (1939); *Cact. Amer.* 14:110 (1942), J. R. Brown; *The First Fifty Haw.* 39 (1970), Pilbeam) has shorter leaves, less narrowly pointed towards the tip; it is distinctly slower-growing than most, and makes an extremely attractive, tight clump of stems over a long period of time. *H. viscosa* var. *torquata* (Haw.) Bak. (*Suppl. Pl. Succ.* 123 (1819); *Monogr.* 3:6 (1836–63), Salm-Dyck; *Journ. Linn. Soc.* 18:201 (1880), Baker; *Cact. Amer.* 14:143 (1942), J. R. Brown; *The Second Fifty Haw.* 59 (1975), Pilbeam) has longer leaves than any other variety, which twist on the stem, making for a most appealing appearance, especially when a clump has developed, which for this variety takes some time. *H. viscosa* var. *viridissima* G. G. Smith (*Journ. S.A. Bot.* 11:67 (1945); *Cact. Amer.* 34:109 (1962), J. R. Brown) is a variety with more intensely green leaves than others, which tend to be coloured more of an olive-green or brown in full light. Finally *H. viscosa* var. *viscosa*, the type, whose identity, as with so many variable species, has become uncertain, but can be taken to be synonymous with the less extreme varieties described above.

Linn. Soc. 18:200 (1880), Baker), similarly has straight rows of leaves, more ovate-triangular than the preceding variety, which has lanceolate-triangular leaves; again this hardly warrants any recognition for separation. *H. viscosa* var. *pseudotortuosa* (Salm-Dyck) Bak. (*Cat. Rais.* 8:1817; *Suppl Pl. Succ.* 59 (1819),

H. viscosa fa. *pseudotortuosa*

H. viscosa fa. *pseudotortuosa*

H. viscosa fa. *subobtusa*

H. viscosa fa. *subobtusa*

On the related species Bayer has little to say except to reduce them to synonymy. But *H. asperiuscula* represents a thick-leaved form of *H. viscosa*, and with varieties which can be related to field populations (G. G. Smith's *H. asperiuscula* var. *subintegra* and var. *patagiata* described in *Journ. S.A. Bot* 11:68 (1945) & 12:11 (1946)) is worth retaining at form level—see below. *H. cordifolia* seems to have been at most a longer-leaved form of the foregoing, and is discarded as insufficiently identifiable. *H. beanii* is a slow, low-growing form, with a more open loose-growing habit than most forms of *H. viscosa*, and with longer leaves than most, to over 4cm long, but without the height of *H. viscosa* var. *torquata*; it is maintained below as a form of *H. viscosa*.

The species resolves as follows:

H. viscosa (L.) Haw. fa. *viscosa* (syn. *H. viscosa* var. *caespitosa*, *H. viscosa* var. *cougaensis*, *H. viscosa* var. *concinna*, *H. viscosa* var. *indurata*, *H. viscosa* var. *quaggaensis*, *H. viscosa* var. *viridissima*).

H. viscosa fa. *asperiuscula* (Haw.) Pilbeam, stat. nov. (syn. *H. asperiuscula* Haw., *H. asperiuscula* var. *patagiata* G. G. Smith, *H. asperiuscula* var. *subintegra* G. G. Smith, *H. cordifolia* Haw.).

H. viscosa fa. *beanii* (G. G. Smith) Pilbeam, comb. nov. (syn. *H. beanii* G. G. Smith, *H. beanii* var. *minor* G. G. Smith, *Journ. S.A. Bot* 10:137 & 138 (1944)).

H. viscosa fa. *pseudotortuosa* (Salm-Dyck) Pilbeam, stat. nov. (syn. *Aloe pseudotortuosa* Salm-Dyck, *H. pseudotortuosa* (Salm-Dyck) Haw., *H. viscosa* var. *pseudotortuosa* (Salm-Dyck) Baker).

H. viscosa fa. *subobtusa* (von Poelln.) Pilbeam, stat. nov. (syn. *H. viscosa* var. *subobtusa* von Poelln.).

H. viscosa fa. *torquata* (Haw.) Pilbeam, stat. nov. (syn. *H. torquata* Haw., *H. viscosa* var. *torquata* (Haw.) Bak.).

The above forms represent the more distinct variations of the species, as already described, worth seeking out from a collector's point of view. But this is a widespread species, and no doubt intergrading forms may be found.

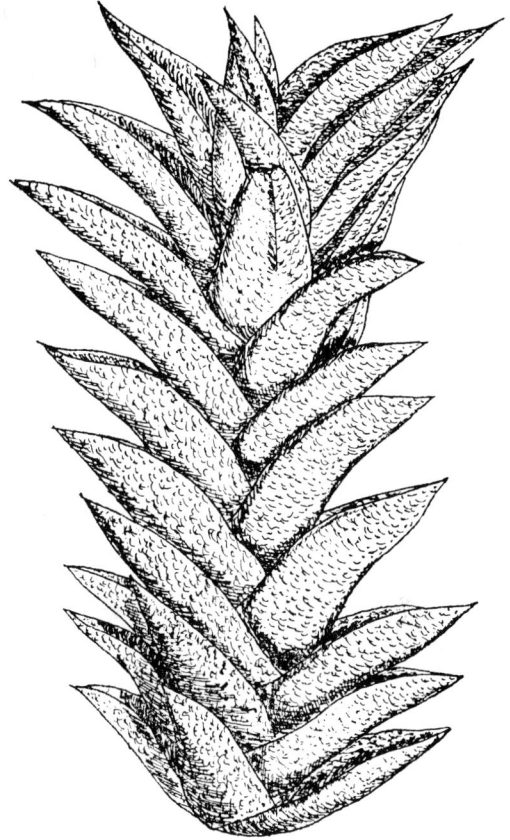

H. viscosa fa. *torquata*

H. viscosa fa. *torquata*

H. viscosa fa. *viscosa*

H. vittata Bak., *Saund. Refug. Bot.* 4:t.263 (1871); *Cact. Amer.* 46:166 (1974), Bayer & Pilbeam; *Haw. Handb.* 167 (1976), Bayer

A vigorous, heavily clumping, light green form of *H. cooperi*.

H. willowmorensis von Poelln., *Cact. Journ.* 5:33 (1936); *Fedde's Repert. Spec. Nov.* 41:216 (1937), von Poellnitz; *Aloe* 11:42 (1973), Scott; & 12:95 (1975), Bayer; *Excelsa* 5:83 (1975); & 7:37 (1977), Bayer

The determination of this species is surrounded with difficulties, and it is regarded by Bayer as probably synonymous with *H. mirabilis*.

H. wittebergensis Barker, *Journ. S.A. Bot.* 8:245 (1942); *Ashingtonia* 1:59 (1974), Pilbeam; *The Second Fifty Haw.* 60 (1975), Pilbeam
Section Loratae subsection Loratae

This species has a hard time of it in the wild, but seems singularly ungrateful when it is brought into intensive care in our collections. It grows in rock fissures in northern faces (the sunny side in the southern hemisphere, of course) of the Witteberg mountain at about 1300m altitude, in deposits of sandy soil, very slowly, shedding its leaf-tips as it suffers from drought, purple-green and with the leaves incurving and closely bunched for protection. In

cultivation it is grey-green, the leaves nearly upright, recurving in the upper half, the margins lined with small teeth, also present in the upper part of the leaf-keel. It is difficult to grow, and care with watering and a gritty compost are needed.

Reported from the Witteberg mountain in the Laingsburg division of Cape Province.

H. wittebergensis

H. wittebergensis

H. woolleyi

H. woolleyi, von Poelln., *Fedde's Repert. Spec. Nov.* 42:269 (1937); *The Second Fifty Haw.* 61 (1975), Pilbeam; *Haw. Handb.* 168 (1976), Bayer; *New Haw. Handb.* 79 (1982), Bayer
Section Venosae

This is a deceptively very slow-growing species. It has the overall appearance of something akin to *H. venosa* subsp. *tessellata*, but is thinner-leaved, and with more the upright habit of *H. attenuata*. The leaves are always dark, almost black-green, with darker edges, toothed on the margins. It will slowly form a small clump.

Reported from the Springbokvlakte area of east of Steytlerville. Little is known of the extent of its distribution, as few collections have been made.

H. xiphiophylla Baker, *Fl. Cap.* 6:345 (1896) Dyer; *Haw. Handb.* 168 (1976), Bayer
Section Arachnoideae subsection Arachnoideae

Although regarded by von Poellnitz as a variety of *H. setata*, this has been restored to specific status by Bayer. It is a long, narrow-leaved species, with little translucence in the leaves, which are sparsely clothed with large prominent teeth on the margins and keel.

It is reported from Sandfontein, Uitenhage.

H. woolleyi

H. xiphiophylla

H. zantnerana von Poelln., *Fedde's Repert. Spec. Nov.* 41:217 (1937); *Haw. Handb.* 169 (1976), Bayer; *New Haw. Handb.* 61 (1982), Bayer
Section Loratae subsection Loratae

This is a pale-green-leaved species with lighter markings longitudinally along the leaves, and a paler margin, colouring to lilac shades in full light. Although clumping and growing well for some time, it will quite readily respond badly to overwatering and die right back, shrivelling alarmingly, before rooting down and getting going again. The leaves are long-triangular and soft in texture, with little rigidity. It is reported from Campherspoort near Klipplaat. Bayer suggests that where it occurs north of Willowmore and in the Baviaanskloof it may be continuous with *H. divergens*.

H. zantnerana

ASTROLOBA

ASTROLOBA

A book on *Haworthia* would not be complete without some reference to the genus *Astroloba*, the most closely related of the other genera in the family Liliaceae. As indicated in the introduction to the chapter on Classification Astrolobas are regarded as distinct, although they have from time to time been joined with *Haworthia*, most recently by Parr in the journal of the African Succulent Plant Society (6:145 (1971)).

For the sake of completeness, and since there are so few species, they are covered in this separate chapter.

In passing it should be mentioned that the monotypic genus *Poellnitzia* has been associated with *Astroloba* and thence with *Haworthia*. The species *Poellnitzia rubriflora*, with, as the specific name indicates, red flowers, has far closer allegiance with the genus *Aloe*, and Gorden Rowley has recently united it with that genus, see fig. on page 154.

Cultivation

The remarks on cultivation for Haworthias apply equally to Astrolobas, although if anything Astrolobas seem to be mainly species which tolerate willingly more arid conditions, and it is difficult to keep them in prime condition in cultivation, especially without their indulging in the unsightly habit of drying up odd leaves on the tight spirals of leaves they make. Various causes for this have been put forward: keeping them too dry, physically damaging the hard tips of the leaves, drying prior to the emergence of a new shoot; but none seem fully to explain this nasty habit.

Classification

With so few species they hardly need classifying, and no attempt is made here to define the minor differences which would warrant separation into groups. The flowers differentiate them from Haworthias of course, and comparison of the picture of a typical Astroloba flower with those of the Haworthias shows the distinctly formed flower tube and arrangement of the petals, which gave the genus its name, *Astroloba* meaning 'with star-shaped lobes to the flowers', i.e. their regular placement as compared with the flared, three-up and three-down arrangement in Haworthias. The nearest that Haworthia flowers come to those of

Astroloba is in the subgenus *Robustipedunculares*, where the arrangement is more regular, but still mighty different from Astroloba.

Flowers of genus *Astroloba*

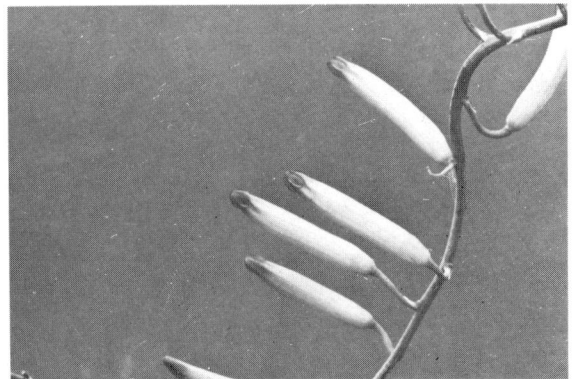

Flowers of *Aloe (Poellnitzia) rubriflora*

Distribution

The distribution of Astroloba species is within the same area as many Haworthias occur, although there is little evidence of hybridization between the two, with one notable exception, *Astroloba skinneri* (see below). The reported localities for each species are given within the commentary on species below.

Checklist of Species

The following list of accepted species is amplified in
the section below 'Commentary on species', but serves
as a check list.

Astroloba aspera var. *aspera*
Astroloba aspera var. *major*

Astroloba bicarinata

Astroloba congesta

Astroloba deltoidea

Astroloba dodsoniana

Astroloba foliolosa

Astroloba herrei

Astroloba pentagona

Astroloba spiralis

Commentary on species

Little work has been done on rationalizing the species
and varieties erected for this genus, apart from
Uitewaal's sorting out of the names at the time he set
up the genus in 1947 to replace the invalid name
Apicra. Since then there have been only two new
species described, and others, not yet published, referred to in an unpublished paper by Mrs P. Roberts-
Reynecke of Cape Town University as a thesis. This
paper covers the genus admirably, and suggests a few
names for species, which in that author's opinion are
indeterminate; these include *A. smutsiana*, *A. rugosa*
and *A. hallii*, in substitution for the older *A. pentagona*
and *A. aspera*.

A. aspera (Haw.) Uitew., *Succulenta* 53 (1947);
Trans. Linn. Soc. 7:7 (1804), Haw.—*Aloe aspera*;
Berlin. Mag. 5:274 (1811), Willdenow; *Syn. Pl. Succ.*
90 (1812), Haw—*Haworthia aspera*; *Monogr.* 3:2
(1836–63), Salm-Dyck—*Aloe aspera*; *Cact. Amer.*
12:19 (1940), J. R. Brown—*Apicra aspera*, *Rev. Pl.*
Succ. 63 (1819), Haworth—*Apicra aspera* var. *major*

This species is well-known in cultivation as a
sprawling, narrow-stemmed plant, with leaves rough
to the touch from tubercling covering the surface. It is
light green when growing strongly becoming red if
exposed to full light. Its rate of growth is slow,
although the sprawling stems may get to about 20cm

A. aspera var. *aspera*

A. aspera var. *aspera*

after a few years, offsetting from around the base and up the stem. The stems of the type are about 2 to 3cm overall in width. The larger-growing *A. aspera* var. *major* has stems 5 or 6cm in diameter and grows more upright.

Reported from Springbokkeel and in the vicinity of Montagu, and near Waterford in the Eastern Cape.

A bicarinata (Haw.) Uitew., *Succulenta* 53 (1947); *Suppl. Pl. Succ.* 63 (1819), Haw.—*Apicra bicarinata*; *Nat. Cact. Succ. Journ.* 28:7 (1973), Rowley in error

Controversy has raged over this species involving author after author, but the painting contemporary with Haworth's description in the collection at the Herbarium Library, Kew, shows clearly a plant redescribed in 1930 by von Poellnitz as *Apicra egregia*. This older name of *A. bicarinata* is taken in preference to the later *A. egregia*.

This species forms stems to 20cm or more tall, collapsing under their own weight after about 15cm, reluctantly offsetting in later life. The leaves are obliquely curved towards their tips, the backs rounded and keeled at an angle, with dark green tubercles, irregularly arranged, the keel and margins with rows of small teeth; the double keel for which the species is named is not always evident, and when present is often only at the lower part of the keel.

Reported from a locality near Oudtshoorn and in the Ceres Sutherland Karoo.

A. aspera var. *major*

A. bicarinata

A. bullulata (Jacq.) Uitew., *Succulenta* 53 (1947); *Fragm.* t.109 (1809) Jacquin—*Aloe bullulata*; *Berlin. Mag.* 5:273 (1811), Willdenow—*Apicra bullulata*

There seems little doubt that if this name is to be applied anywhere it is to the natural hybrid between *A. aspera* and *Haworthia pumila*, generally seen labelled as *A. skinneri*. Since this is the prior name for this intergeneric hybrid it takes preference over Rowley's *Astroworthia bicarinata* nm. *skinneri*.

A. congesta (Salm-Dyck) Uitew., *Succulenta* 54 (1947); *Monogr.* 2:1 (1836–63), Salm-Dyck—*Aloe congesta*; *Journ. Linn. Soc.* 18:218 (1880), Baker—*Apicra congesta*; *Journ. Bot.* 27:44 (1889)—*Apicra turgida*

The massive stem of this species depicted in Salm-Dyck's Monograph is indicative of the size it can reach, up to 25cm or more tall and nearly 10cm in overall width. The plant pictured here was imported and first on arrival it was about 15 cm tall and about 7 cm wide; as shown it is the rooted top of the old stem, which had lost many of its lower leaves through getting excessively dry at the root. The leaves are dark green,

hard and smooth in texture, with sharp points and no tubercling. The margins and keel have a cartilaginous edge and there is a tendency to concavity in the upper part of the back of the leaf and the upper surface, especially if kept a little on the dry side.

A. deltoidea (Hook f.) Uitew., *Succulenta* 54 (1947); *Bot. Mag.* t.6071 (1873), Hook fils—*Aloe deltoidea*; *Journ. Linn. Soc.* 18:217 (1880), Baker—*Apicra deltoidea*; *Cact. Amer.* 12:60 (1940), J. R. Brown—*Apicra deltoidea*

The standing of this species is questionable, and fieldwork may well reveal that it is simply a smaller-growing *A. congesta*. It forms tall, eventually sprawling stems to about 20cm or more long, remaining upright until they reach about 15cm or so, offsetting from around the base. The width is about 4 or 5cm overall. The leaves are dark green, smooth, shining and rigid with a sharp point. The varieties *intermedia* and *turgida* are not considered worth upholding, being merely variations in size or degree of spiralling, which in the typical *A. deltoidea* is slight. Reported from stony places in the Zuurberg mountains.

A. congesta

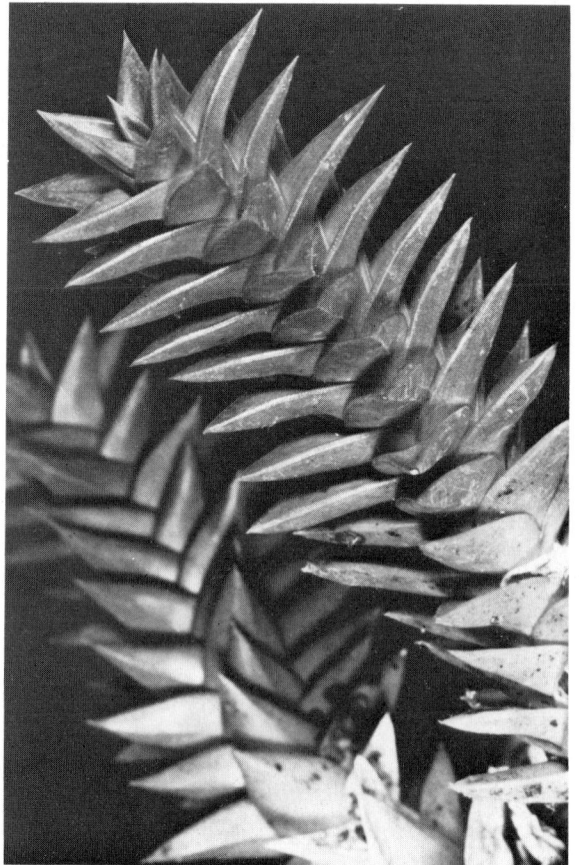

A. deltoidea

A. dodsoniana Uitew., *Des. Pl. Life* 22:29 (1950)

Although it has been suggested that this species may be a larger-growing form of *A. herrei*, it is sufficiently different to warrant its retention for collectors, and until its relationship is determined by field examination I am happy to retain it in any consideration of the genus. It makes long, sprawling stems to about 25cm or more long and about 4 to 6cm wide overall. The leaves are smooth, inflated below, tapering to a sharp point, and an overall grey-green enhanced by a light covering of a waxy nature, in full light making the plants appear white. It offsets in time from around the base.

Reported from Uniondale and Prince Albert.

A. dodsoniana

A. egregia (von Poelln.) Uitew., *Succulenta* 54 (1947); *Fedde's Repert. Spec. Nov.* 28:100 (1930), von Poellnitz—*Apicra egregia*; *Des. Pl. Life* 9:33 (1937), von Poellnitz; *Cact. Amer.* 12:3 (1940), J. R. Brown—*Apicra egregia*

There is little doubt after looking at the painting in the collection of the Herbarium Library, Kew, contemporary with Haworth's description of *A. bicarinata,* that this is a later description of that species. Uitewaal's *A. egregia* var. *fardeniana* differed only in the spiral arrangement of the leaves compared with the alleged straight, longitudinal rows in the type. In fact it is a question of the degree of the spiralling, which all Astroloba species do to a greater or lesser degree.

A. foliolosa (Haw.) Uitew., *Succulenta* 54 (1947); *Trans. Linn. Soc.* 7:7(1804), Haworth—*Aloe foliolosa*; *Berlin Mag.* 5:274 (1811)—*Apicra foliolosa*; *Monogr.* 2:4 (1836–63), Salm-Dyck; *Fl. Cap.* 6:331 (1896) Dyer, Baker—*Apicra foliolosa*; *Cact. Amer.* 12:44 (1940), J. R. Brown—*Apicra foliolosa*

Roberts-Reynecke in her unpublished thesis takes the view that this is the preferred name for a number of synonymous species, including *A. deltoidea* and *A.*

congesta, proposing retention of the latter at subspecific level. In absence of publication the latter name is retained at specific level here, although the probable synonymy of *A. deltoidea* is acknowledged.

The species forms an extremely tightly spiralling, narrow, but rigid column of leaves, only collapsing when it has reached probably its maximum height of about 20cm. The leaves stand out almost at a rightangle from the stem, and are flat-triangular, rigid and extremely sharply pointed. They are coloured usually light to mid-green turning to purplish-brown in full light.

Reported from the Ladismith district, from between Ladismith and Laingsburg; the Steytlerville district, from Steytlerville; the Uitenhage district, from Addo Bush, Kleinpoort, Wolvefontein; the Jansenville district, from Waterford; the Graaff Reinet district, from Graaff Reinet, Kruidfontein, Kendrew.

A. foliolosa

A. hallii n.n., P. Roberts-Reynecke, unpublished thesis on *Astroloba*

This is the preferred name adopted by Mrs Roberts-Reynecke for *A. pentagona*, which she regards as confused and indeterminate. Since this is the type species of the genus it should not be lightly discarded, and the author thinks that there is enough evidence in the literature to indicate that *A. pentagona* is identifiable as grown in collections today. See under *A. pentagona*. The plant identified by Mrs Roberts-Reynecke with the new name *A. hallii* is an attractive species, as yet unnamed in my opinion, that is to say not conspecific with any previously described species as suggested by Mrs Roberts- Reynecke. It remains to be validated, and this book is not the place where that should be done. It is nevertheless pictured for interest, as it is an attractive collector's plant, slow-growing, grey-green in colour, with a milky-white quality, with whitish-green cartilaginous margins and oblique keel, and irregularly scattered, similarly coloured, low tubercles beneath the surface.

A. hallii n.n.

A, herrei Uitew., *Des. Pl. Life* 20:37 (1948); *Succulenta* 56 (1949–50), Uitewaal

This is a charming, small-growing, small-leaved species, with stems only about 4cm wide overall, getting to about 20cm long, sprawling and branching from the base and along the stem, especially as it lowers from the upright position. Leaves are light, glaucous-green, tapering to a long sharp point, with many darker green longitudinal lines beneath the surface, the margins tending to colour reddish-brown in full light.

Reported from Uniondale and Prince Albert.

A. pentagona (Ait.) Uitew., *Succulenta* 54 (1947); *Hort. Kew.* 1:471 (1789), Aiton—*Aloe spiralis* var. *pentagona*; *Spec. Pl.* (1799), Willdenow—*Aloe spiralis* var. *pentagona*; *Trans. Linn. Soc.* 7:7 (1804), Haworth—*Aloe pentagona*; *Berlin. Mag.* 5:273 (1811),

A. herrei

Willdenow—*Aloe pentagona*; *Fragm.* t.III (1809), Jacquin—*Aloe pentagona*; *Monogr.* 1:4 (1836–63), Salm-Dyck—*Aloe pentagona*; *Cact. Amer.* 15:31 (1943), J. R. Brown—*Apicra pentagona*

This is the type species of the genus, and there is little doubt after examining the literature of the late eighteenth and early nineteenth centuries that this species can be equated to plants in cultivation today. The stems are erect and grow to about 20cm or more, 3 to 4cm wide overall, clustering from the base. Leaves are ovate-lanceolate, sharply pointed, green reddening in full light, smooth, not shining, with an oblique keel and no tubercles.

Varieties ascribed to this species in the past are discounted as being based on differences too trivial to warrant consideration. They include *A. pentagona* var. *spiralis* (most plants under this name are believed to be of hybrid origin, with possibly *Haworthia* as one of the parents; they cluster prolifically, grow quickly, have tubercles beneath the surface, and have a somewhat softer texture than most *Astroloba* species, and the flowers are not wholly characteristic of the genus), *A. pentagona* var. *spirella* and *A. pentagona* var. *torulosa*; there is also var. *quinquangularis* of the previously specifically ranked *Aloe spirella*.

A. skinneri (Berg.) Uitew., *Succulenta* 53 (1947); *Pflanz.* 4 (38): 116 (1908), Berger—*Apicra skinneri*; *Nat. Cact. Succ. Journ.* 28:7 (1973), Rowley—*Astroworthia bicarinata* nm. *skinneri*

That this is a naturally occuring hybrid there is no doubt, and its parentage seems to have been established as *Astroloba aspera* × *Haworthia pumila*. Unfortunately it has been reduced beneath *Astroworthia* (former *Astroloba*) *bicarinata*, on the false supposition, in my opinion, that it too is a hybrid, while, as I have explained above under that name, I believe that evidence in the Herbarium Library, Kew, refutes this. It is therefore raised, if that is the word, to the rank of an F1 hybrid as *Astroworthia skinneri*, stat. nov. It is wide-stemmed (to 7 or 8cm at the base) tapering, about 10 to 15cm tall in maturity, with spiralling, broad, ovate-lanceolate leaves, tubercles on the lower surface, and with usually two keels, both keels, margins and tubercles a paler green than the matt mid-green of the leaves, turning red in full light.

A. spiralis (L.) Uitew., *Succulenta* 53 (1947); *Sp. Pl.* 322 (1753), Linnaeus—*Aloe spiralis*; *Trans. Linn. Soc.* 7:7 (1804)—*Aloe imbricata*; *Fragm.* 226.t.110 (1809), Jacquin—*Aloe spiralis*; *Monogr.* 1:1 (1836–63), Salm-Dyck—*Aloe imbricata*; *Cact. Amer.* 11:83 (1939), J. R. Brown

A. pentagona

A. spiralis

This is a wonderfully spiralling species, well-named for its closely overlapping, neatly arranged leaves. The stems are 20cm or more tall, and about 3cm wide overall, clustering usually from around the base only, and reaching 15cm or more before sprawling. The leaves are ovate-lanceolate, tapering gradually to a sharp point, and with a distinct oblique keel.

A. turgida (Bak.) Jacobs., *Handb. Succ. Pl.* 229 (1960); *Journ. Bot.* 27:44 (1889), Baker—*Apicra turgida*; *Succulenta* 54 (1947), Uitewaal

Uitewaal quite rightly referred this species to the prior, if larger *A. congesta*. It is not regarded as significantly different enough to warrant recognition.

A. spiralis

Aloe (Poellnitzia) rubriflora

GLOSSARY OF TERMS

In order to assist the reader to pronounce and understand the names for Haworthias, the following glossary has been prepared, particularly geared to Haworthia.

Many Haworthia species' names are made up of the name of a person or a place with a suffix: -ensis, as in *H. wittebergensis*, which indicates that this species comes from Witteberg; -iana or -ana, as in *H. comptoniana* or *H. zantnerana* (after 'r' or a vowel—including 'y'—the 'i' of -iana is dropped), meaning that these species are named in honour of Compton and Zantner; -ii (or -i after 'r' or a vowel, including 'y'), -iae (or -ae after 'r' or a vowel, including 'y'), as in *H. reinwardtii*, *H. herrei*, *H. woolleyi*, *H. blackburniae*, *H. emelyae*, named respectively after the finders of the species, Reinwardt, Herre, Woolley (male, hence the ending -i), and Blackburn and Emely (Ferguson) (female, hence -ae); note that either surname or first name may be used.

Other than place or people's names for species, the name is usually descriptive in some way of the species, e.g. *H. truncata*, meaning truncated, referring to the leaves, which have the appearance of having had the ends cut off, or *H. graminifolia*, with grass-like leaves.

Pronunciation is a vexed topic, and not so important as some would have us believe. Listen to any foreign enthusiast's way of pronouncing the names, and you will soon find anomalies brought about by attempting to use a dead language as a universal tongue, when different tongues have different ways of interpreting the Latin or Greek construction, which is the basis of their meaning or make-up.

Accent falls usually on the penultimate syllable if it is a 'long' syllable, or if the name has only two syllables. Otherwise it falls on the syllable before the penultimate one, e.g. *H. coarctāta*, but *H. decīpiens*.

The vowels and consonants which most often present difficulty are explained below, but there are, as with all 'rules', exceptions:

ae = ee as in keep
au = a as in ball
c is hard (k' before a, o and u, and soft (s) before e, i and y
ch is either hard (k) or soft, as in chimney
ei = i as in bite
g is hard, as in game, before a, o and u, and soft, as in gentle, before e, i and y
oe = ee as in keep
ui is pronounced as two syllables, not as in suit.

a-	not or without
acantha	spine or prickle
acaul-	without, or apparently without, a stem
acuminata	tapering to a long point
aegrota	unwell, sick-looking
affinis	related or having affinity to
agavoides	like an Agave
albi-	white
albicans	becoming white
albinota	marked with white
-alis	pertaining to
alti-	tall or long
ambigua	ambiguous, doubtful
angusti-	narrow
apicra	not sharp or bitter
arachnoid	cobweb-like
aranea	with cobwebs
argenteo	silvery
aristata	with a long end-bristle
armata	armed with bristles
aspera, -ula, -iuscula	rough, roughened
astro-	star-like
atro-	dark
attenuata	tapering, long and slender
awn	bristle
baccata	berry-like
badia	dark reddish-brown
bellula	beautiful
bi-	two, twice
bracteata	with bracts, modified leaves, usually on the flower stem
'brevicula	short
brunnea	brown
bullulata	blistered
caespitosa	offsetting, growing in clumps
carinata	keeled
cassytha	like a genus of tropical vines
caulescentes	producing stems
chloro-	green or yellowish-green

clari-	bright or clear
coarctata	crowded
color	coloured
columnaris	column-like
concinna	elegant, neat
confluens	blending into one
congesta	crowded together, congested
conspicua	conspicuous, notable
cordi-	heart-shaped
correcta	corrected
crassa	thick
crystallina	ice-like, crystal-like
cultivar, cv.	a form of plant originating in cultivation
curta	short
cuspidata	sharp-pointed
cymbiformis	boat-like
decipiens	deceiving, deceptive
delineata	lined, delineated
deltoidea	triangular
densi-	dense, close-set
dentata	toothed
denticulifera	densely toothed
diminuta	diminutive, very small
dimorpha	having two forms
divergens	diverging from a central point
diversi-	differing, distinct
egregia	excellent
erecta	upright, erect
falcata	curved, sickle-like
fallax	deceptive
fasciata	forming a crest, joined together
fenestrata	windowed
ferox	fierce
floribunda	with abundant flowers
folia	leaves
foliolosa	with many leaves
form, forma, fa.	plant or group of plants below specific level and below variety level, not sufficiently distinct to justify varietal status
formis	formed
fulva	tawny-yellow
fuscus	brown, or grey-brown
genus	taxonomic category containing species with some characters in common
gigantea	very large, gigantic
gigas	giant
glabrata	smooth
glauca	bluish-grey or bluish-green, with waxy bloom
globosi-	globose, spherical
gracili-, gracilis	graceful, slender
gramini-	grass-like
grandis, grandicula	large, grand

granulata	covered with small grains, granulate, roughened
guttata	spotted, with translucent markings
habdomadis	having seven parts (this species was found at Sevenweekspoort)
herbacea	herbaceous, not woody
hexangulares	hexangular, with six sides
hybrida	a hybrid
icosi-	twenty
imbricata	overlapping
inconfluens	in (river's) confluence
incurvula	curving inwards
indurata	hardened
integra	entire
intermedia	intermediate
-iorum	genitive plural ending (of two or more)
-issima	superlative, most
laete	bright
laevis	smooth, polished
lanceolata	lance-shaped
lepida	scaly
liliputana	very small
limi-	file-like
limpida, -ula	clear, translucent
linea, lineata	lined or striped
loba	lobe or segment, round-edged
loratae	strap-shaped
luteo	yellow
maculata, -osa	spotted
magnifica	magnificent
major	larger, greater
margaritifera	having pearls
marginata	with a distinct border or edge
minima	smallest
minor	smaller
mirabilis	wonderful, marvellous
montana	of mountains
monticola	from the mountains
mucronata	ending in a short, sharp point
multi-	many
mundula	clean, neat or elegant
musculina	like a little mouse, or, like a muscle
mutabilis	changeable
mutica	blunt, without points
nana	small, dwarf
nigra	black
nigricans	becoming black
nitidula	shining
notabilis	notable, distinct
obesa	fat, swollen
obtusa	blunt, rounded
-oides	similar to, with the appearance of
olivacea	olive-green
ovato	egg-shaped, with the broader end downwards
pallida, pallidi-	pallid, pale
papillosa	with small, blunt projections or warts

paradoxa	strange, seemingly contradictory
pauci-	few
pedunculares	pertaining to the stalks of flowers
pentagona	five-angled
per-	entirely
perplexa	perplexing, anomalous
phylla	of a leaf
picta	painted, patterned
pilifera	very hairy
plani-	flat
prolifera	offsetting freely
pseudo-	false, like
pubescens	covered with fine, soft, short hairs
pulchella	small and beautiful
pulchra	beautiful
pumila	dwarf, small
pygmaea	very small
quin- quangularis	five-angled
radula	rasp-like
ramifera	very branching
recurva	recurving, curving downwards or backwards
regularis	regular, uniform
reticulata	having a network (of veins in a leaf)
retusa	with a rounded apex, lightly notched
rigida	stiff, inflexible, rigid
robusti-	strong, thick, robust
rosea	rose-coloured, pink
rubro-	red
rugosa	roughened, wrinkled
salina	salty
scabra	rough
semi-	half
serrata	like a saw, toothed
setata	bristly
sessili-	without a stalk
solitaria	solitary, single
sordida	dirty, muddy
sparsa	scattered, sparse
species	a population of individuals which breeds true within its own limits of variations, and shows distinct discontinuity from other species
spiralis, spirella	spiralled
stolonifera	offsetting by stolons, stems produced from the base of the plant, above or below ground, producing offsets at the tips
striata	marked with lines or ridges
sub-	somewhat, nearly, partially
subspecies	taxonomic category, below species, but above variety; plants forming the subspecies have distinguishing features from others of the same species, but are not sufficiently distinct to be given species status
subulata	long and tapering to a stiff point
taxon (plural taxa)	a name in taxonomy for any one of any rank
taxonomy	the study of classification of organisms
tenera	delicate, soft
tenuis	slender, thin
tessellata	marked with small squares, like a mosaic
torquata	wearing a collar
tortella	twisted
tortuosa	very twisted
torulosa	cylindrical with bulges or contractions at intervals
transiens	transient, passing (found in the Prince Albert Pass)
translucens	partially transparent, translucent
trifariae	with three rows
truncata	cut off at the tip
tuberculata	with tubercles, small warty protuberances
turgida	full of sap, and so rigid
umbraticola	coming from the shade
uni-	one, single
valida	well developed
variabilis	variable, varying
variegata	particoloured, with patches of different colouring
variety, var.	taxonomic category within a species; the plants have distinguishing features from others of the same species but are not sufficiently distinct to be given species status or subspecies status
venosa	with veins
virens	green
virescens	becoming green
viridis	green
viridissimus	very green
viscosa	sticky
vittata	with longitudinal stripes or lines
viva	living
xiphio-	sword, sword-like
zebrina	striped, like a zebra

SOCIETIES

At the time of writing there are the following societies devoted to furthering the knowledge of cacti and succulent plants. None is concerned specifically with the genus *Haworthia*, but some are more orientated towards the succulent, as opposed to cactus, side of the hobby, and contain perhaps more in-depth study of these plants.

Great Britain

The National Cactus & Succulent Society
Membership Secretary: Miss W. E. Dunn
43 Dewar Drive
Sheffield

Providing more or less equally for cacti and succulents in the quarterly journal. Has over 100 branches throughout the UK, which usually meet monthly. Seed distribution annually.
Basic annual subscription (1982) £5.
Amalgamation with the following society in active discussion at the time of going to press.

The Cactus & Succulent Society of Great Britain
67 Gloucester Court
Kew Road
Richmond
Surrey

Similarly equally divided between the cacti and succulent sides of the hobby in its three or four journals a year. Seed distribution annually. Amalgamation with the preceding society being actively pursued at the time of going to press. The amalgamation will probably result in the continuation of a quarterly journal, plus the production of a year-book, optional to members for an extra fee, in which more weighty articles will be collected together. Basic annual subscription (1982) £6.50.

The Xerophyte
Barleyfield
Southburgh
Thetford
Norfolk

A small journal devoted entirely to the succulent side of the hobby, published quarterly. Seed distribution annually. Subscription (1982) £3.

Zimbabwe

The Aloe, Cactus & Succulent Society of Zimbabwe (formerly Rhodesia)
PO Box 8514
Causeway
Salisbury
Zimbabwe

Publishes a year-book, with the accent on succulent plants, entitled *Excelsa*. There have been several lengthy articles on Haworthia in the last few years. Subscription ZR5.

South Africa

The South African Aloe & Succulent Society
PO Box 1193
Pretoria 0001
South Africa

Publishes an irregular journal entitled *Aloe*, mainly on succulent plants. Subscription SAR10. There have been problems resulting in the irregular issue of journals in the last few years, but past issues have contained lengthy articles on Haworthia.

Australia

The Cactus & Succulent Society of New South Wales
32 High Street
Woonoona
New South Wales 2517
Australia

This society's journal has contained findings of a study group on Haworthias over a period of several years, with good accompanying photographs. Subscription (1982) A$6.

United States of America

The Cactus & Succulent Society of America, Inc.
Abbey Garden Press
1675 Las Canoas Road
Santa Barbara
California 93105, USA

More or less equally divided between the two sides of the hobby, with perhaps a slight inclination towards the more popular native cacti of the Americas. Publishes six journals a year. Basic annual subscription (1982) $16. Past issues, from the 1930s until the early 1970s, contained an invaluable series of '*Notes on Haworthias*' by J. R. Brown.

West Germany

Kakteen und andere Sukkulenten
Deutsche Kakteen-Gesellschaft e.V.
Moorkamp 22
D-3008 Garbsen 5
West Germany

Publishes 12 journals a year (in German of course), with more accent on cacti. Basic annual subscription (1982) DM34, joining fee DM8.

Those detailed above are the more prominent societies in the hobby, but there are others, most of which publish journals, in Canada, France, Mexico, India, and elsewhere.

BIBLIOGRAPHY

Aloe — *Aloe*, 1963– (journal of the South African Aloe and Succulent Society), Bayer, Scott

Ashingtonia — *Ashingtonia*, 1974 (journal of the Reference Collection, Ashington), Pilbeam

ASPS — *African Succulent Plant Society Journal*

Beitr. Sukk. — *Beiträge zur Sukkulentenkunde und -Pflege*, 1939–40 von Poellnitz

Berl. Deutsch. Bot. Ges. — *Berlin, Deutsche Botanische Gesellschaft*, Marloth, 1909, Schönland

Berlin Mag. — *Hortus Berolinensis*, 1811, Willdenow

Bol. Soc. Brot. — *Boletim de Sociedade Broteriana*, 1940–41, Resende

Cact. Amer. — *Journal of the Cactus & Succulent Society of America*, 1929– J. R. Brown, Scott, Bayer, Pilbeam

Cact. GB. — *Journal of the Cactus & Succulent Society of Great Britain* (formerly *Cactus Journal*), 1932–82, von Poellnitz

Cact. Journ. — *Cactus Journal* (London) 1898–1900

Cact. Succ. Journ. NSW. — *Journal of the Cactus & Succulent Society of New South Wales*, 1971–74, Bayer

Cact. Vetpl. — *Cactussen en Vetplanten*, 1939–40, Uitewaal

Cat. Rais. — *Catalogue Raisonné des Espèces d'Aloes*, Salm-Dyck, 1817

Des. Pl. Life — *Desert Plant Life*, 1929–43, von Poellnitz, J. R. Brown

Encycl. — *Encylopédie Méthodique*, 1783, 1797, Lamarck

Excelsa — *Excelsa*, 1971– (journal of the Aloe, Cactus & Succulent Society of Zimbabwe – formerly Rhodesia)

Fedde's Repert. Spec. Nov. — *Repertorium Specierum Novarum Regni Vegetabilis*, Fedde, 1929–40, von Poellnitz

The First Fifty Haw. — *The First Fifty Haworthias*, 1970, Pilbeam

Fl. Cap. — *Flora Capensis*, 1896, Baker

Fl. Pl. Afr./Fl. Pl. S.Afr. — *Flowering Plants of Africa/Flowering Plants of South Africa*, 1940, 1941, 1946

Fragm. — *Fragmenta*, 1809, Jacquin

Gard. Dict. — *Gardener's Dictionary*, 1768, Miller

Haw. Handb. — *Haworthia Handbook*, 1976, Bayer

Hort. Kew — *Hortus Kewensis*, 1789, Aiton

Hort. Dyck. — *Hortus Dyckensis*, 1834, Salm-Dyck

Hort. Med. Amst — *Horti Medici Amstelodamensis*, 1701, Commelin

Hort. Schoenb. — *Hortus Schönbrunensis*, 1804, Jacquin

Journ. Bot. — *Journal of Botany*, 1929, Bolus

Journ. Linn. Soc. — *Journal of the Linnaean Society*, 1880, Baker

Journ. S.A. Bot. — *Journal of South African Botany*, 1935– , (various authors, notably G. G. Smith in the 1940s)

Kakt. u. and. Sukk. — *Kakteen und andere Sukkulenten*, 1950– (journal of the West German Cactus & Succulent Society)

Kakteenk. — *Kakteenkunde*, 1937, 1938, von Poellnitz

Kew. Bull. — *Kew Bulletin*

Mem. Soc. Brot. — *Memorias da Sociedade Broteriana*, 1941–43, Resende

Monogr. — *Monographia Generum Aloes et Mesembryanthemi* (1836–63), Salm-Dyck

Nat. Cact. Succ. Journ. — *Journal of the National Cactus & Succulent Society* (UK), 1946–82, Uitew., Bayer

New Haw. Handb. *The New Haworthia Handbook*, 1982, Bayer

Notizbl. Bot. Gart. Mus. Berl. *Notizblatt des Botanischen Garten und Museums zu Berlin-Dahlem*, 1906, Marloth & Berger

Obs. Bot. *Observationes Botanicae*, 1820, Salm-Dyck

Pflanz. *Das Pflanzenreich*, Engler, 1908, Berger

Phil. Mag. *Philosophical Magazine*, 1824–26, Haworth

Pl. Succ. Hort. Alenc. *Plantae Succulentae in Horto Alenconio*, 1809, Duval (reproduced in the *Cactus Journal* 8:105 (1939))

Pl. Succ. Hort. Dyck. *Plantae Succulentae Hortus Dyckensis* 1816, Salm-Dyck

Port. Acta Biol. *Portugaliae Acta Bioligica*, 1946, 1948, Resende & Viveiros

Rec. Albany Mus. *Records of the Albany Museum*, 1912, Schönland

Refug. Bot. *Refugium Botanicum*, Saunders, 1870 et seq., Baker

Revis. Pl. Succ. *Revisiones Plantarum Succulentarum*, Haworth, 1821

The Second Fifty Haw. *The Second Fifty Haworthias*, 1975, Pilbeam

Sp. Pl. *Specierum Plantarum*, 1753, Linnaeus

Spec. Pl. *Species Plantarum*, 1799, Willdenow

Succulenta *Succulenta*, 1947–53, Uitewaal (journal of the Dutch Cactus & Succulent Society)

Sukkulentenk. *Sukkulentenkunde*, 1951

Suppl. Pl. Succ. *Supplementum Plantarum Succulentarum*, Haworth, 1819

Syn. Pl. Succ. *Synopsis Plantarum Succulentarum*, Haworth, 1812

Trans. Linn. Soc. *Transactions of the Linnaean Society*, 1804, 1880, Haworth, Baker

Trans. Roy. Soc. S. Afr. *Transactions of the Royal Society of South Africa*, 1908, 1910, 1932, Marloth, Fourcade

INDEX

Acaules   15, 121, 123, 124
Aloe   9, 72, 147
  aspera   100, 149
  asperula   118
  atrovirens   81
  bullulata   150
  congesta   150
  cylindrica var. *rigida*   120
  deltoidea   150
  foliolosa   151
  herbacea   81
  humilis   72
  imbricata   153
  margaritifera   105
  mirabilis   96
  pentagona   153
  pseudotortuosa   141
  pumila var. *margaritifera*   92
  reticulata   114
  rubriflora   154
  skinneri   123
  spiralis   153, 154
  spirella   153
  turgida   154
Apicra   14, 148
  aspera   148
  aspera var. *major*   148
  bicarinata   149
  bullulata   150
  congesta   150
  deltoidea   150
  egregia   149, 151
  foliolosa   151
  pentagona   152, 153
  skinneri   153
  spiralis var. *pentagona*   152
  turgida   150, 154
Arachnoideae   12, 14, 25, 35, 37, 38, 45, 48, 57, 61, 67, 78, 94, 122, 129, 143
Ariocarpus kotschoubeyanus   103
Astroloba   9, 14, 74, 100, 147, 151
  aspera   100, 123, 148, 150
  aspera x *H. pumila*   153
  aspera var. *aspera*   148
  aspera var. *major*   148, 149

  bicarinata   148, 149, 151, 153
  bullulata   150
  congesta   148, 150, 151, 154
  deltoidea   148, 150, 151
  deltoidea var. *intermedia*   150
  deltoidea var. *turgida*   150
  dodsoniana   148, 151
  egregia   149, 151
  egregia var. *fardeniana*   151
  foliolosa   148, 151
  hallii   148, 151
  herrei   148, 151, 152
  pentagona   148, 152
  pentagona var. *quinquangularis*   153
  pentagona var. *spiralis*   153
  pentagona var. *spirella*   153
  pentagona var. *torulosa*   153
  rugosa   148
  skinneri   147, 150, 153
  smutsiana   148
  spiralis   148, 153
Astroworthia bicarinata   153
  bicarinata nm. *skinneri*   123, 150, 153
  skinneri   153
Attenuatae   15, 41, 88, 107

Calibanus hookeri   78
Caulescentes   15, 100, 138
Ceropegia armandii   124
Chortolirion   14
Coarctatae   15, 27, 39, 41, 51, 71, 74, 88, 107, 108
Cultivation   11
Cymbifoliae   15, 61

Fenestraria   131
Fenestratae   11, 15, 94, 131
Frithia   131
Fusiformes   14, 15, 46, 77

Gasteria   9. 13
  batesiana   124
Growing medium   11

Haworthia aegrota   35, 81
  affinis   46

agavoides   123, 124
albanensis   36
albicans   92
altilinea   14, 35
altilinea denticulata   35
altilinea var. *limpida* fa. *inconfluens*   78, 82
altilinea var. *morrisiae*   78
angustifolia   15, 29, 35, 36, 46, 50, 68, 82, 99, 104, 105
angustifolia fa. *angustifolia*   36
angustifolia fa. *baylissii*   29, 36
angustifolia fa. *grandis*   36
angustifolia fa. *paucifolia*   36
angustifolia var. *albanensis*   36
angustifolia var. *denticulifera*   36, 50, 51
angustifolia var. *liliputana*   36, 50, 51
angustifolia var. *subfalcata*   36
arachnoidea   14, 21, 28, 29, 37, 38, 82, 91, 105, 122, 123
aranea   14, 29, 37, 38
archeri   15, 38
archeri var. *archeri*   38
archeri var. *dimorpha*   38
aristata   14, 28, 29, 38, 39, 72, 80, 95, 120, 134, 138
aristata var. *aristata*   39
aristata var. *helmiae*   39
armstongii   15, 39, 40, 41, 49
aspera   149
asperiuscula   39, 138, 140, 141
asperiuscula var. *patagiata*   140, 141
asperiuscula var. *subintegra*   140, 141
asperula   41, 107, 118
atrofusca   41, 89
attenuata   15, 29, 41, 42, 43, 71, 84, 86, 88, 107, 108, 109, 113, 120, 127, 143
attenuata fa. *attenuata*   41, 42
attenuata fa. *britteniae*   45
attenuata fa. *britteniana*   42, 45
attenuata fa. *caespitosa*   13, 42, 43, 44, 45, 71
attenuata fa. *caespitosa* (variegated form)   45
attenuata fa. *clariperla*   42, 43, 45
attenuata var. *caespitosa*   41
attenuata var. *clariperla*   42
baccata   45
badia   45, 96
batesiana   14, 21, 29, 45, 46
batteniae   46
baylissii   36, 46
beanii   46, 138, 140, 141
beanii var. *minor*   46, 141
bilineata   46
bilineata var. *gracilidelineata*   46, 61
blackbeardiana   46,57
blackburniae   15, 28, 29, 46, 47, 77, 155
bolusii   14, 21, 37, 38, 46, 48, 57, 122
bolusii var. *aranea*   49
bolusii var. *blackbeardiana*   21, 49
bolusii var. *bolusii*   48
bolusii var. *semiviva*   49, 122
browniana   49, 71
bruynsii   29, 49
caespitosa   49, 132, 133
carrissoi   50
cassytha   50
chalwinii   52
chloracantha   15, 28, 36, 50, 68
chloracantha var. *chloracantha*   51

chlorancantha var. *denticulifera*   36, 51
chloracantha var. *subglauca*   51
coarctata   15, 29, 45, 51, 52, 55, 71, 74, 78, 81, 99, 103, 109, 111, 112, 150
coarctata subsp. *adelaidensis*   51, 52, 55, 111
coarctata subsp. *adelaidensis* fa. *adelaidensis*   50
coarctata subsp. *adelaidensis* fa. *bellula*   52, 55
coarctata subsp. *coarctata* var. *coarctata*   51
coarctata subsp. *coarctata* var. *coarctata* fa. *coarctata*   52, 53
coarctata subsp. *coarctata* var. *coarctata* fa. *chalwinii*   52, 54
coarctata subsp. *coarctata* var. *coarctata* fa. *conspicua*   52, 55
coarctata subsp. *coarctata* var. *greenii*   51, 52
coarctata subsp. *coarctata* var. *tenuis*   51, 52, 113
coarctatoides   55
comptoniana   15, 29, 57, 118, 155
comptoniana fa. *comptoniana*   56, 57
comptoniana fa. *major*   10, 57
cooperi   14, 21, 29, 35, 57, 66, 84, 102, 104, 142
cooperi var. *cooperi*   58
cooperi var. *cooperi* fa. *cooperi*   57, 58, 59
cooperi var. *cooperi* fa. *pilifera*   13, 58, 59, 103, 125
cooperi var. *leightoniae*   58
cooperi var. *leightonii*   18, 57, 58, 60
cordifolia   60, 138, 140, 141
correcta   48, 60, 69
cuspidata   60, 91
cymbiformis   15, 21, 29, 46, 60, 61, 77, 82, 84, 94, 102, 104, 108, 134
cymbiformis var. *cymbiformis*   61, 62
cymbiformis var. *cymbiformis* (*brevifolia*)   62
cymbiformis var. *cymbiformis* (*compacta*)   62
cymbiformis var. *cymbiformis* fa. *cymbiformis*   13, 61
cymbiformis var. *cymbiformis* fa. *cymbiformis* (variegated form)   13, 62
cymbiformis var. *cymbiformis* fa. *gracilidelineata*   61, 62, 63, 66
cymbiformis var. *cymbiformis* fa. *multifolia*   61, 62, 63
cymbiformis var. *cymbiformis* fa. *obesa*   61, 64, 65
cymbiformis var. *cymbiformis* fa. *planifolia*   61, 65
cymbiformis var. *cymbiformis* fa. *ramosa*   61, 64, 65
cymbiformis var. *incurvula*   46, 61, 66
cymbiformis var. *multifolia*   61
cymbiformis var. *obesa*   61
cymbiformis var. *transiens*   29, 61, 66
cymbiformis var. *translucens*   61, 66
cymbiformis var. *umbraticola*   61, 66, 67
decipiens   14, 29, 67, 68, 87, 104, 155
dekenahii   68, 117
dekenahii var. *argenteo-maculosa*   68, 117
divergens   15, 29, 68, 144
eilyae   69, 75
eilyae var. *zantnerana*   69
emelyae   15, 28, 29, 41, 57, 60, 69, 90, 96, 98, 104, 107, 118
emelyae var. *emelyae*   69, 70, 71
emelyae var. *beukmannii*   96, 98
emelyae var. *multifolia*   69, 71
fasciata   15, 29, 41, 42, 43, 49, 71, 113
fasciata fa. *browniana*   72
fasciata fa. *fasciata*   19, 72
fasciata fa. *major*   71
fasciata fa. *ovato-lanceolata*   72
fasciata fa. *perviridis*   71

fasciata fa. *sparsa* 72, 73
fasciata fa. *subconfluens* 72
fasciata fa. *vanstaadensis* 72
fasciata fa. *variabilis* 72
fasciata var. *concolor* 72
fasciata var. *major* 72
fasciata var. *perviridis* 72
ferox 72
floribunda 15, 28, 73, 74, 103
floribunda x *H. retusa* fa. *longebracteata* 74
fouchei 74, 119
fulva 74
geraldi 74, 116, 119
glabrata 15, 74, 82, 88, 128
glauca 15, 50, 69, 74, 82
glauca var. *glauca* 29, 74, 75
glauca var. *herrei* 29, 39, 69, 74, 75
glauca var. *herrei* fa. *herrei* 76
glauca var. *herrei* fa. *jacobseniana* 74, 75, 77
glauca var. *herrei* fa. *jonesiae* 74, 75, 77
globosiflora 77, 101
gracilidelineata 61, 77
gracilis 77
graminifolia 15, 29, 46, 77, 78, 155
granulata 78, 136
greenii 51, 78
guttata 78
haageana 78, 114, 116
haageana var. *subreticulata* 78, 114
habdomadis 14, 78, 82, 99
habdomadis var. *habdomadis* 28, 78, 79
habdomadis var. *inconfluens* 28, 78, 79
habdomadis var. *morrisiae* 29, 78, 80
heidelbergensis 15, 28, 69, 80, 81, 96
helmiae 80, 134
henriquesii 81
herbacea 15, 28, 35, 72, 81, 82, 88, 89, 91, 103, 105, 114, 128, 132
herrei 82, 155
hurlingii 82, 114
hurlingii var. *ambigua* 114
hybrida 82
icosiphylla 82
inconfluens 78, 82
incurvula 82
integra 82
intermedia 82
isabellae 82
jacobseniana 74, 82
janseana 82
jonesiae 75, 82
cv. *kewensis* 82, 83
kingiana 15, 28, 83, 123
koelmaniorum 15, 21, 83, 85
krausiana 84
krausii 84
cv. *kuentzii* 83, 84
laetevirens 84, 103
lateganiae 84, 124, 132, 133
leightoniae 57
leightonii 57, 84
lepida 62, 84
limifolia 15, 21, 83, 84, 86, 87, 91, 134
limifolia var. *gigantea* 86, 87
limifolia var. *keithii* 86, 87

limifolia var. *limifolia* 84, 87
limifolia var. *limifolia* fa. *limifolia* 84
limifolia var. *limifolia* fa. *major* 84, 86, 87
limifolia var. *schuldtiana* 86
limifolia var. *stolonifera* 86
limifolia var. *stolonifera* fa. *major* 87
limifolia var. *stolonifera* fa. *pimentelii* 87
limifolia var. *striata* 85, 87
limifolia var. *ubomboensis* 87, 133
lisbonensis 87
lockwoodii 14, 29, 79, 87, 88, 122
longebracteata 88, 119
longiana 5, 29, 88
longiana var. *albinota* 88
longibracteata 88
luteorosea 88
maculata 15, 28, 88, 128
magnifica 15, 28, 41, 78, 89, 91, 96, 103, 105, 118, 122, 123
magnifica var. *atrofusca* 28, 89, 90
magnifica var. *magnifica* 89
magnifica var. *major* 28, 71, 89, 90
magnifica var. *maraisii* 28, 36, 78, 89, 90, 105, 118
magnifica var. *meiringii* 89, 90, 127
magnifica var. *notabilis* 89, 91
magnifica var. *paradoxa* 28, 72, 89, 91
cv. *mantelii* 91, 92
maraisii 89, 90, 92
maraisii var. *meiringii* 127
margaritifera 92, 95, 103, 106, 122
marginata 15, 28, 92, 93, 106, 134
marginata var. *laevis* 92
marginata var. *ramifera* 92
marginata var. *virescens* 92
marginata x *H. minima* 93
marumiana 14, 21, 29, 38, 45, 94
maughanii 15, 29, 94, 95, 124
mclarenii 95
minima 15, 20, 28, 95, 96, 99, 104, 106, 128, 134
mirabilis 15, 28, 45, 90, 91, 96, 99, 101, 120, 131, 132, 142
mirabilis subsp. *badia* 98, 99
mirabilis subsp. *mirabilis* fa. *beukmannii* 98
mirabilis subsp. *mirabilis* fa. *mirabilis* 97
mirabilis subsp. *mirabilis* fa. *napierensis* 98
mirabilis subsp. *mirabilis* fa. *rubrodentata* 97
mirabilis subsp. *mirabilis* fa. *sublineata* 97, 98
mirabilis subsp. *mundula* 99, 103
monticola 99
morrisiae 99, 121
mucronata 18, 99
mucronata var. *morrisiae* 78, 99
mucronata var. *mucronata* fa. *inconfluens* 18, 82, 99
mundula 96, 99
musculina 99
mutabilis 99
mutica 15, 28, 96, 99, 100, 103, 116, 118
nigra 15, 21, 29, 100
nigra fa. *angustata* 101
nigra fa. *nana* 101
nigra fa. *nigra* 100
nigra var. *angustata* 101
nigra var. *diversifolia* fa. *nana* 101
nortieri 15, 21, 77, 101
nortieri var. *giftbergensis* 101
nortieri var. *globosiflora* 21, 101
nortieri var. *montana* 101

nortieri var. *nortieri* 101
notabilis 89, 91, 102
obtusa 57, 102, 104
cv. *ollasonii* 102
otzenii 99, 103
pallida 103
pallida var. *paynei* 81, 103
papillosa 103, 106
papillosa var. *semipapillosa* 103, 106
paradoxa 89, 91, 103, 118
parksiana 15, 28, 74, 103
peacockii 83, 103
pearsonii 104
pellucens 66
perplexa 104
picta 69, 104
pilifera 57, 104
pilifera var. *columnaris* 104
pilifera var. *dielsiana* 104
pilifera var. *gordoniana* 104
pilifera var. *salina* 104
pilifera var. *stayneri* 104
planifolia 61, 66, 104
planifolia var. *transiens* 61, 104
poellnitziana 15, 28, 104
pseudogranulata 105
pseudotessellata 136
pseudotortuosa 141
pubescens 15, 28, 105, 118
pulchella 15, 28, 105
pumila 15, 28, 83, 92, 95, 103, 104, 105, 106, 122, 123, 127, 150
pygmaea 15, 28, 51, 100, 106, 118
pygmaea fa. *crystallina* 10, 107
pygmaea fa. *major* 10, 106, 107
pygmaea fa. *pygmaea* 106
radula 15, 29, 107, 108, 120, 128
ramosa 61, 108
recurva 108, 128, 136
reinwardtii 13, 15, 29, 51, 52, 99, 108, 109, 111, 112, 113, 155
reinwardtii var. *adelaidensis* 109, 111
reinwardtii var. *archibaldiae* 111
reinwardtii var. *bellula* 52, 109, 111
reinwardtii var. *brevicula* 109, 111, 112, 113
reinwardtii var. *chalumnensis* 109, 111
reinwardtii var. *chalwinii* 52, 111
reinwardtii var. *committeesensis* 112
reinwardtii var. *conspicua* 52, 112
reinwardtii var. *diminuta* 109, 111, 112
reinwardtii var. *fallax* 112
reinwardtii var. *grandicula* 110, 112
reinwardtii var. *huntsdriftensis* 112
reinwardtii var. *kaffirdriftensis* 109, 112
reinwardtii var. *major* 111
reinwardtii var. *minor* 111
reinwardtii var. *olivacea* 109, 113
reinwardtii var. *peddiensis* 110, 113
reinwardtii var. *pulchra* 113
reinwardtii var. *reinwardtii* fa. *chalumnensis* 113
reinwardtii var. *reinwardtii* fa. *kaffirdriftensis* 113
reinwardtii var. *reinwardtii* fa. *olivacea* 113
reinwardtii var. *reinwardtii* fa. *reinwardtii* 113
reinwardtii var. *reinwardtii* fa. *zebrina* 113
reinwardtii var. *riebeekensis* 113

reinwardtii var. *tenuis* 113
reinwardtii var. *triebneri* 113
reinwardtii var. *valida* 113
reinwardtii var. *zebrina* 113
resendeana 113, 114
reticulata 15, 28, 78, 82, 89, 91, 114, 132
reticulata var. *acuminata* 114
reticulata var. *hurlingii* 115, 116
reticulata var. *reticulata* 114
reticulata var. *subregularis* 115, 116
retusa 15, 28, 60, 68, 74, 88, 103, 116, 117, 119, 132
retusa var. *acuminata* 18, 89, 118, 119
retusa var. *argenteo-maculosa* 117
retusa var. *dekenahii* 117, 119
retusa var. *densiflora* 116, 117, 118
retusa var. *multilineata* 116, 117, 119
retusa var. *retusa* fa. *fouchei* 117, 118, 119
retusa var. *retusa* fa. *geraldii* 118, 119
retusa var. *retusa* fa. *longebracteata* 118, 119
retusa var. *retusa* fa. *multilineata* 119
retusa var. *retusa* fa. *retusa* 116, 119
retusa var. *solitaria* 116, 117
revendettii 119, 120, 121
rigida 87, 120
rossouwii 96, 97, 120
rubrobrunnea 120
rugosa 120
rugosa var. *perviridis* 120
rycroftiana 15, 120
ryderana 120
sampaiana 120, 121
sampaiana fa. *broterana* 121
scabra 15, 28, 29, 99, 121, 128, 132
scabra var. *morrisiae* 29, 121, 122
scabra var. *scabra* 20, 121
schmidtiana var. *angustata* 101
schuldtiana 89, 118, 122
schuldtiana var. *erecta* 78
schuldtiana var. *maculata* 122
schuldtiana var. *major* 89, 90, 122
semiglabrata 122
semiviva 14, 21, 122
serrata 15, 28, 69, 122, 123
sessiliflora 123
setata 37, 123, 143
setata var. *bijliana* 37
setata var. *gigas* 37
setata var. *joubertii* 37
setata var. *nigricans* 37
setata var. *xiphiophylla* 37
skinneri 123
smitii 15, 123
sordida 15, 29, 123, 124
sordida var. *agavoides* 123, 124
sordida var. *lavrani* 123, 124
springbokvlakensis 15, 29, 49, 124
starkiana 15, 29, 84, 123, 124
starkiana var. *lateganiae* 29, 125
starkiana var. *starkiana* 124, 125
stiemei 125
cv. *subattenuata* 13, 126, 127
cv. *subattenuata* (variegated form) 126, 127
subfasciata 83, 127
subfasciata var. *kingiana* 127
sublimpidula 89, 127

submaculata 81, 128
subulata 128
tauteae 128
tenera 128, 129
tisleyi 128
torquata 141
tortuosa 128, 129
tortuosa (variegated form) 129
tortuosa var. *curta* 128
tortuosa var. *major* 129
tortuosa var. *pseudorigida* 128
tortuosa var. *tortella* 128
translucens 14, 29, 66, 77, 82, 125, 128, 129
translucens subsp. *tenera* 11, 66, 95, 129
translucens subsp. *translucens* 129
triebnerana 97, 131
triebnerana var. *napierensis* 98
triebnerana var. *rubrodentata* 97
triebnerana var. *turgida* 98
truncata 15, 29, 91, 95, 124, 131, 155
truncata fa. *crassa* 130, 131
truncata fa. *tenuis* 131, 132
truncata fa. *truncata* 19, 130, 131
truncata var. *tenuis* 131, 132
truncata x *H. limifolia* 132
turgida 15, 28, 29, 49, 60, 68, 69, 72, 82, 84, 103, 117, 120, 132
turgida fa. *caespitosa* 133
turgida fa. *pallidifolia* 133
turgida fa. *suberecta* 133
turgida fa. *turgida* 132, 133
turgida var. *pallidifolia* 132, 133
turgida var. *suberecta* 132, 133
turgida var. *subtuberculata* 132, 133
ubomboensis 15, 87, 133, 134
uitewaaliana 92, 134
umbraticola 61, 134
unicolor 38, 39, 120, 134
variegata 15, 28, 68, 134
venosa 15, 21, 28, 78, 91, 108, 128, 136
venosa subsp. *granulata* 136
venosa subsp. *recurva* 136
venosa subsp. *tessellata* 21, 28, 29, 105, 128, 136
venosa subsp. *tessellata* *(parva)* 136
venosa subsp. *venosa* 135, 136
venosa var. *oertendahlii* 136
venteri 39, 134, 138
viscosa 15, 21, 28, 29, 39, 46, 128, 129, 138, 140
viscosa fa. *asperiuscula* 139, 141
viscosa fa. *beanii* 139, 141

viscosa fa. *pseudotortuosa* 140, 141
viscosa fa. *subobtusa* 140, 141
viscosa fa. *torquata* 141
viscosa fa. *viscosa* 20, 140, 141
viscosa var. *caespitosa* 138, 141
viscosa var. *coegaensis* 138
viscosa var. *concinna* 138, 141
viscosa var. *cougaensis* 138, 141
viscosa var. *indurata* 138, 141
viscosa var. *pseudotortuosa* 139, 141
viscosa var. *quaggaensis* 139, 141
viscosa var. *subobtusa* 139, 141
viscosa var. *torquata* 139, 141
viscosa var. *viridissima* 139, 141
viscosa var. *viscosa* 139
vittata 57, 58, 59, 142
willowmorensis 96, 97, 142
wittebergensis 15, 29, 47, 142, 155
woolleyi 15, 29, 143, 155
xiphiophylla 14, 19, 29, 125, 143
zantnerana 15, 29, 68, 144, 155
Hexangulares 15, 16, 19, 20, 21

Limpidae 14, 35, 48, 57, 78, 122, 123, 129
Lithops 131
Loratae 15, 25, 35, 46, 50, 68, 73, 77, 103, 105, 134, 142, 144
Maps 22–27
Margaritiferae 15, 83, 92, 95, 104, 106, 123

Neogomesia 123

Pests 13
Poellnitzia 14, 147
Poellnitzia rubriflora 14, 147, 154
Proliferae 14, 45, 94
Propagation 12

Retusae 11, 15, 25, 38, 57, 80, 81, 88, 89, 96, 99, 101, 103, 105, 114, 116, 120, 122, 123, 124, 132
Robustipedunculares 15, 16, 17, 18, 20, 27, 83, 134, 147

Shading 11

Trifariae 15, 27, 100, 121, 123, 124, 138
Turgidae 15, 38, 81, 88, 101, 105, 114, 120, 122, 132

Variegation 12, 45, 63, 126, 127, 130
Venosae 15, 27, 83, 84, 133, 136, 143

Watering 11